11th Southeast Asia Astronomy Networking Meeting (SEAAN Meeting 2019)

EPJ Web of Conferences Volume 240 (2020)

Singapore
26 - 28 December 2019

Editors:

W.K. Ng
A. Yang
S.C.C. Ng
A.H. Chan
C.H. Sow

ISBN: 978-1-7138-1796-3

Printed from e-media with permission by:

Curran Associates, Inc.
57 Morehouse Lane
Red Hook, NY 12571

Some format issues inherent in the e-media version may also appear in this print version.

No changes have been made to the content of these proceedings. There may be changes to pagination and minor adjustments for aesthetics.

Printed with permission by Curran Associates, Inc. (2021)

For additional information, please contact EDP Sciences – Web of Conferences at the address below.

EDP Sciences – Web of Conferences
17, Avenue du Hoggar
Parc d'Activité de Courtabœuf
BP 112
F-91944 Les Ulis Cedex A
France

Phone: +33 (0) 1 69 18 75 75
Fax: +33 (0) 1 69 28 84 91

contact-edps@webofconferences.org

Additional copies of this publication are available from:

Curran Associates, Inc.
57 Morehouse Lane
Red Hook, NY 12571 USA
Phone: 845-758-0400
Fax: 845-758-2633
Email: curran@proceedings.com
Web: www.proceedings.com

TABLE OF CONTENTS

Statement of Peer review

In submitting conference proceedings to *Web of Conferences*, the editors of the proceedings certify to the Publisher that

1. They adhere to its **Policy on Publishing Integrity** in order to safeguard good scientific practice in publishing.
2. All articles have been subjected to peer review administered by the proceedings editors.
3. Reviews have been conducted by expert referees, who have been requested to provide unbiased and constructive comments aimed, whenever possible, at improving the work.
4. Proceedings editors have taken all reasonable steps to ensure the quality of the materials they publish and their decision to accept or reject a paper for publication has been based only on the merits of the work and the relevance to the journal.

Title, date and place of the conference

The 11th Southeast Asia Astronomy Networking Meeting (SEAAN Meeting 2019)
26 to 28 December 2019
Singapore

Proceedings editor(s):

Ng, W.K., Yang, A., Ng, S.C.C., Chan, A.H., Sow, C.H.

Date and editor's signature

06 August 2020

Foreword

Local Organising Committee, SEAAN Meeting 2019

The National University of Singapore was honoured to host the 11th Southeast Asia Astronomy Network Meeting in conjunction with an annular solar eclipse in December 2019.

The Southeast Asia Astronomy Network (SEAAN) Meeting was first held in 2007 during the Thai National Astronomy Meeting in 2007, hosted by the National Astronomy Research Institute of Thailand (NARIT). Since then the SEAAN Meeting has rotated among Southeast Asian countries. Recent Meetings were held in Bandar Lampung (Indonesia, 2018), Mandalay (Myanmar, 2017) and Hanoi (Vietnam, 2016).

The purpose of the SEAAN Meeting is to bring together people working in the fields of astronomy and astrophysics from Southeast Asia. It aims to facilitate exchanges in both research findings and educational practices, as well as astronomy outreach efforts. The 2019 SEAAN Meeting consisted of plenary talks and poster sessions over two days with over 70 participants from various ASEAN countries and beyond.

We acknowledge the generous sponsorship from the NUS physics department towards this conference. Special thanks go to the members of the Local Organising Committee, in particular Maggie Lau, Samuel Wu and Ben Kek whose efforts ensured the success of the meeting.

We also acknowledge the efforts of the SEAAN chair and vice-chair Professors Boonrucksar Soonthornthum and Hakim Malasan respectively, as well as the previous hosts for building up the SEAAN meeting to what it is today. We record our thanks to the scientific organising committee whose work in reviewing abstracts and manuscripts have been key to the success of the SEAAN Meetings.

Picture 1: Conference Participants

Picture 2: Business Meeting

Picture 3: Left: Solar eclipse viewing event; Right: Ring of Fire at maximum eclipse

In conjunction with the meeting, there was an annular solar eclipse visible from Singapore on 26 December 2019. This was an extremely rare event and a fitting occasion to celebrate astronomy in Southeast Asia.

Local Organising Committee

- Dr. Abel Yang (chair)
- Dr. Cindy Ng (co-chair)
- Dr. Ng Wei Khim (proceedings)
- Mr. Samuel Wu (admin)
- Ms. Maggie Lau (admin)
- Assoc. Prof. Phil Chan (deputy head, NUS Physics)
- Prof. Sow Chorng Haur (head, NUS Physics)

Scientific Organising Committee

- Prof. Boonrucksar Soonthornthum (chair)
  Thailand
- Assoc. Prof. Hakim L. Malasan (vice chair)
  Indonesia
- Dr. Chey Chan Oeurn
  Cambodia
- Prof. Hasan Abu Kassim
  Malaysia
- Prof. Khin Swe Myint
  Myanmar
- Dr. Cynthia Celebre
  Philippines
- Dr. Pham Ngoc Diep
  Vietnam
- Dr. Abel Yang
  Singapore

List of Participants

Name	Affiliation
Dr. Abel Yang	National University of Singapore
Mr Alfred Tan	National Junior College, Singapore
Mr Ang Han Wei	National University of Singapore
Dr. Aprilia	Institut Teknologi Bandung
Prof. Boonrucksar Soonthornthum	National Astronomical Research Institute of Thailand
Ms. Cao Shangyu	National Junior College-School of Science and Technology, Singapore
Dr. Chan Oeurn Chey	Royal University of Phnom Penh
Prof. Chelsea Sharon	Yale-NUS College
Mr. Chen-Fatt Lim	National Taiwan University / Academia Sinica Institute of Astronomy & Astrophysics

Name	Affiliation
Dr. Ching Chee Leong	Singapore University of Technology and Design
Dr. Cindy Ng	National University of Singapore
Engr. Dario L. Dela Cruz	Philippine Atmospheric, Geophysical and Astronomical Services Administration
Mrs. Darunee Lingling Orchiston	
Dr. David Mkrtichian	National Astronomical Research Institute of Thailand
Dr. Do Thi Hoai	Vietnam National Space Center
Dr. Eddie Chua	Institute of High Performance Computing, A*STAR
Mrs. Evaria Puspitaningrum	Xtremax Ltd.
Dr. Evgenii Semenko	National Astronomical Research Institute of Thailand
Mr. Fahmi Iman Alfarizki	Institut Teknologi Bandung
Mr. Fargiza A. M. Mulki	Institut Teknologi Bandung
Mr. Gum Ja Naw	Department of Higher Education, Myanmar
Mr. Gupta Arpit	National Junior College-School of Science and Technology, Singapore
Prof. Hakim L. Malasan	Institut Teknologi Bandung
Dr. Hesti Retno T. Wulandari	Institut Teknologi Bandung
Ms. Izatul Hafizah	Institut Teknologi Bandung
Dr. John Soo Yue Han	Universiti Sains Malaysia
Mr. Keith Chia Wen Kai	National Junior College-School of Science and Technology, Singapore
Mr. Koh Joo Beng	
Mr Lim Kian Hwee	National University of Singapore
Dr. Lim Kim Yong	National University of Singapore
Mr. Lim Meng Hwee	Science Centre Board, Singapore
Ms. Lisa Goh Wan Khee	National University of Singapore
Mr. Low Lerh Feng	University of Auckland
Mr. Lu Jianlong	National University of Singapore
Ms. Ma. Rosario C. Ramos	Philippine Atmospheric, Geophysical and Astronomical Services Administration
Mr. Maulana Asgaf	Universitas Gadjah Mada
Mr. Mochammad Dafa Wardana	Institut Teknologi Bandung
Mr. Muhammad Adam Aqasha Bin Mohamed Hisham	National Junior College-School of Science and Technology, Singapore
Mr. Muhammad Nazri B Nordin	National University of Singapore
Ms. Muthia Dewi	Institut Teknologi Bandung
Dr. Ng Wei Khim	National University of Singapore
Mr. Nguyen Tan Vu	Ho Chi Minh City Astronomical Association
Ms. Ni Made Kartika Wijayanti	Institut Teknologi Bandung
Dr. Norhasliza Yusof	Universiti Malaya
Ms. Nur Amira Mohd Norsham	Universiti Teknologi MARA
Ms. Nurul Fathin Ngadiman	Universiti Teknologi MARA
Dr. Oh Boon Kiat	Seoul National University

Name	Affiliation
Mr. Oh Wei Shen	The Australian National University
Mr. Ong Zongjin	National University of Singapore
Dr. Pham Ngoc Diep	Vietnam National Space Center
Dr. Phil Chan	National University of Singapore
Mr. Pulkit Agarwal	National University of Singapore
Ms. Putri Mutmainah Khoirul Anam	Institut Teknologi Bandung
Mr. Quek Zhi Hao	National University of Singapore
Mr. Remus Chua	
Mr. Rizchel M. Masong	De La Salle University
Mr. Rizky Maulana Nurhidayat	Institut Teknologi Bandung
Mr. Seow Kit Hint	National Junior College-School of Science and Technology, Singapore
Mr. Setthawut Thongmee	National Astronomical Research Institute of Thailand
Mr. Shrivastava Kushagra	National Junior College-School of Science and Technology, Singapore
Asst. Prof. Siramas Komonjinda	Chiang Mai University
Ms. Sneha Athreya	National Junior College-School of Science and Technology, Singapore
Dr. Sunly Khimphun	Royal University of Phnom Penh
Mr. Sze-leung Cheung	National Astronomical Research Institute of Thailand
Ms. Taavishi Jindel	Yale-NUS College
Mr. Tan Jyh Harng	Science Centre Board, Singapore
Mr. Toh Kang Hui	Science Centre Board, Singapore
Dr. Wang Wei Yang	National University of Singapore
Prof. Wayne Orchiston	National Astronomical Research Institute of Thailand
Mr. Wei Leong Tee	University of Arizona
Mr. Wichan Insiri	National Astronomical Research Institute of Thailand
Mr. Wong Kang Jun	National Junior College-School of Science and Technology, Singapore
Prof. Yee Yee Oo	Taungoo University

Capturing a Total Solar Eclipse

Joo Beng Koh[1,*]
[1]Institute of Physics, Singapore

Abstract. A Total Solar Eclipse is one of the astronomical world's most awe-inspiring phenomena. It is a relatively rare astronomical event that lasts at most a few minutes and can be viewed only at very specific times and from very specific locations on the globe. Capturing and documenting the event so that it can be shared with others is exceptionally challenging because of its rarity and brevity. This article addresses some of the challenges and possible solutions with the use of various forms of technology as well as program management skills.

1 Total Solar Eclipse

A total solar eclipse is a relatively rare event and only a very small proportion of people actually experience one. More commonly experienced are lunar eclipses and partial solar eclipses, and yet total solar eclipses are not as rare as one may imagine it to be. It happens on average every 1 to 2 years[†], but yet it is so rare because it is only visible from a very narrow path on the Earth and lasts at most 2 to 3 hours. Totality itself, which is the critical period during the eclipse when the Sun is completely obscured by the Moon, is even more fleeting, lasting 2 to 3 minutes on average, and at most about 6 and a half minutes.

1.1 Geometry of Eclipses

Eclipse geometries are determined by the positions of the Earth, the Sun, and the Moon in space. Solar eclipses occur when the Moon is between the Earth and Sun. As the Moon revolves around the Earth about 13 times a year, it may be thought that there would be 13 solar eclipses each year. However, due to the 5° tilt of the Moon-Earth plane relative to the Sun-Earth plane (see Fig. 1), there are only 2 opportunities for a Solar eclipse to occur each year, and can happen only when the Moon is at the intersection of the Moon-Earth plane with the Sun-Earth plane.

With the Moon between the Sun and the Earth, the distance of the Moon from the Earth now comes into play. As the Moon's orbit is slightly elliptical, its distance from the Earth varies as it orbits. This variation in distance results in annular, hybrid, or total eclipses. Total eclipses occur when the Moon is so close to the Earth that the Sun is able to cast an umbral shadow on the Earth's surface; Annular eclipses occur when the Moon is so far away that the Sun only casts an antumbral shadow; and a hybrid eclipse occurs when the Sun casts an

* Email : joobeng@astrobargains.com
† 1991-2000: 7 total eclipses, 2001-2010: 7 total eclipses, 2011-2020: 6 total eclipses, Average: 20 total eclipses in 30 years, or 1.5 years per eclipse.

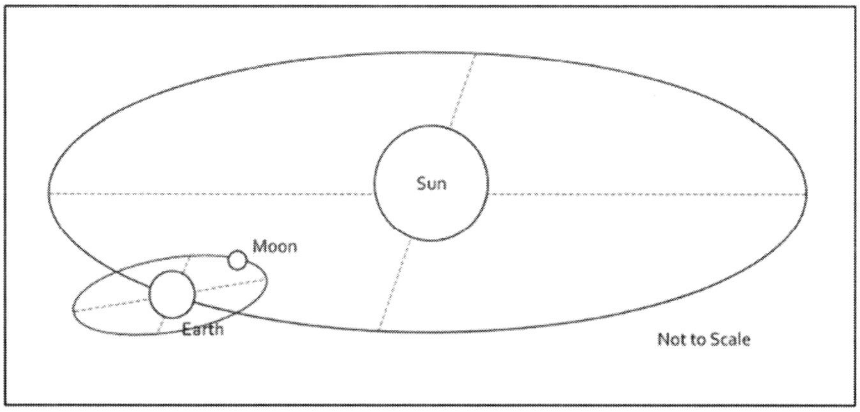

Fig. 1. The Earth-Moon orbit is tilted relative to the Sun-Earth orbit

umbral shadow along a portion of the eclipse path and antumbral shadow for the rest of its path (Fig. 2).

1.1.1 Width of a total eclipse path

The width of an eclipse path is determined by many of the factors discussed so far as well as where the umbral shadow falls upon the Earth. The path will be wider when it falls closer to the edges of the globe (see Fig. 3). The widest that a path on the equator can be (so that we can ignore the curvature of the Earth) is when the Moon is at perigee, the Sun at perihelion, and works out‡ to be just about 100km. Theoretically, if this path is projected at the edges of the globe, it could span about 1000km, so that would be roughly the maximum width.

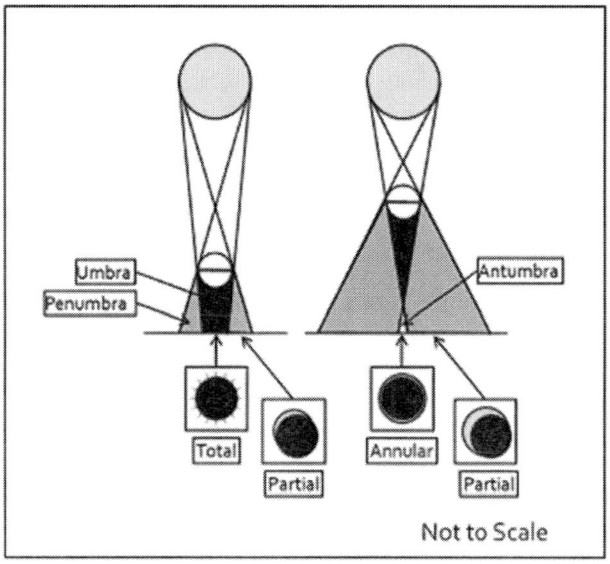

Fig. 2. Types of Solar Eclipses

‡ Assumptions: Distance of Moon at perigee = 356400km, Diameter of Moon = 3474km, Distance of Sun at perihelion = 1.471e8km, Diameter of Sun = 1.391e6km.

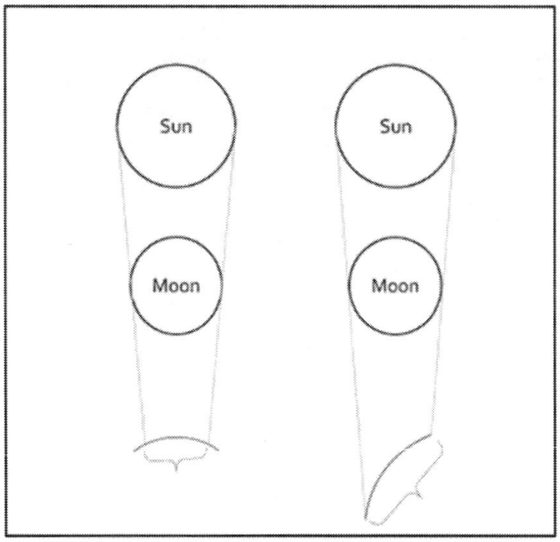

Fig. 3. The path of totality is wider towards the edges of the globe

1.1.2 Duration of a total eclipse

A total eclipse is divided into 3 phases demarcated by 4 events. The 3 phases in sequential order are: a Partial phase, where the Moon is partially covering the Sun; Totality where the Sun is covered by the Moon; and another Partial phase where the Moon is uncovering the Sun. The 4 events are named 1st, 2nd, 3rd, and 4th Contact. The 1st Contact is when the Moon first impinges upon the Sun's disc, signifying the start of the eclipse; 2nd Contact occurs when the Moon completely occludes the Sun, signifying the start of totality; 3rd Contact is when the Moon just uncovers the Sun, signifying the end of totality; and 4th contact is when the Moon leaves the Sun's disc, signifying the end of the eclipse. The duration from 1st to 4th contact typically lasts about 2 to 2.5 hours, and the duration from 2nd to 3rd contact can is at most about 6.5 minutes, with a theoretical minimum of 0 minutes (i.e. the Moon/Sun ratio is 1.0 so the Moon literally covers and leaves the Sun at the same time).

1.2 Eclipse Considerations

To observe an eclipse, the obvious factors are to have the observer at the right place (on the track/path of the eclipse) and at the right time. We shall discuss a few more considerations to be able to observe a total eclipse successfully.

1.2.1 Accessibility of Location.

Unfortunately, an eclipse's path does not always fall on land. In fact, statistically, the odds are that 70% of eclipse paths would be over water[§]. This in and of itself, does not mean that the eclipse would not be observable, as vessels like cruise liners and smaller vessels can always be used as an observation platform, though they are less than ideal due to the motion of the waves. Even when over land though, the area may be difficult to get to, and can include areas such as the Antarctic pole as well as deep in the Amazon jungle.

[§] Ratio of land to sea is 3:7

1.2.2 Safety and Security.

Observation sites are chosen based on the level of safety and security they are able to afford. An obvious choice here would be in or near cities, and where there are no wars or other conflicts occurring. The political alignment of the country is also important to consider. Some countries are also particular about licensing and other regulatory requirements, as permits for travelling, observing, and even importation of equipment for use during the eclipse will need to be addressed.

1.2.3 Weather.

One of the most critical considerations in viewing an eclipse is the weather prospects, and in particular the cloud probability. Good statistical data for cloud cover for the relevant date/time of the year should be considered. Other weather considerations include the temperature, precipitation, and wind, as all these can lead to adverse viewing conditions.

2 Eclipse Oddities

During the phases of a total solar eclipse, there are many observational oddities due to the unique changes in lighting, as well as the optical elements because of the position of the Sun/Moon. These can be sub-divided into 2 main categories.

2.1 Sun/Moon effects

These are effects that can be observed purely from the juxtaposition of the Moon and the Sun.

2.1.1 Diamond Ring and Bailey's Beads

These 2 effects are roughly very similar though Bailey's Beads are for the more "serious" photographers and Diamond Rings are more for the lay public. They occur at 2nd and 3rd contact only, and lasts for a very short time, in the order of seconds or less. The diamond ring is primarily the effect of the glare from the Sun peeping out behind the silhouette of the Moon. Exposure settings are not as critical and is easily captured by most photographers as settings are typically overexposed. Bailey's beads on the other hand, are due to the light from the Sun peeping through the valleys at the Moon's limb, and requires very accurate exposures so that the detail of individual "beads" are not lost. Scientific data can be derived from images of Bailey's beads as they aid in determining or confirming the profile of the lunar surface at the positions of the beads.

2.1.2 Corona

Totality can be viewed with the naked eye without fear of any injury (unlike the rest of the partial phases, where solar filters must be used). And with the naked eye, the Sun's coronal streamers are clearly visible and seem to come alive. These streamers will appear to radiate quite far from the Sun, from 1 to 5 solar diameters. However, capturing the coronal streamers photographically is a challenge because the exposure settings vary quite dramatically depending on how close the streamers are to the Solar Disc. At its closest, the corona is very bright, and exposure settings of 1/8000 sec, f/16, and ISO 800 can be used, whereas further out from the Sun, the corona is dim, and settings such as 16 sec, f/4, and ISO 100 may be required. A common practice is to photograph by bracketing – taking many pictures with a range of exposures. To then convey the coronal streamers and its structure after the eclipse, multiple photos will need to be processed together to extract the relevant information from the photos and reconstruct the coronal streamers that the naked eye so easily sees (because of the human eyes' high dynamic range).

2.1.3 Chromosphere and Prominences

Most think of the Sun as bright white in colour, so during totality, those who see a reddish colour fringing the edges of the Moon think that they are imagining it. This is in fact the chromosphere, which is a thin shell of reddish/pink colour. The colour is due to the transition Hydrogen atoms and occurs just beneath the coronal layer, and is visible only during a total eclipse. Prominences, which appear as pink "flares" emerging from the chromosphere can also be seen. Capturing the chromosphere and prominences are also challenging as a camera's exposure settings have to be low enough so that the light from the rest of the Sun does not wash out the colour and the detail of the chromosphere and prominences.

2.1.4 Face of the Moon

An astute observer may be able to discern that the Moon is not just a black silhouette during totality, and in fact, has some features. Light from the Sun does not directly illuminate the face of the Moon as seen from Earth. It is actually reflected off the surface of the Earth, reflected yet again from the face of the Moon, to finally reach an observer back on Earth. In order to capture this feature, cameras have to be set to very long exposures, yet not too long which may result in light from the Sun blowing out details at the edge of the Moon.

2.2 Terrestrial and other effects

Apart from the Sun and Moon, there are many other observational oddities, both on land as well as in the sky, that can only be seen during a total eclipse. Observers are often so entranced with the sight of totality that they fail to look away and observe the other effects to the sides and behind them.

2.2.1 Sky colour and shape of colour

Daytime sky illumination, which is owing to light rays from the Sun scattering off atmospheric particles, no longer occurs. But this is only true within the umbral shadow of the eclipse. A reduced proportion of light rays are still present in the penumbral shadow, and so too for the light outside of the penumbral shadow, and these rays of light can still reflect off the atmosphere back to an observer in the umbral shadow. The resultant effect is almost similar to twilight, but is also very distinct from it, with colours of the rainbow, not on the horizon, but bowl-shaped. This is because the umbral and penumbral shadows are cone-shaped and when projected through the atmosphere results in a characteristic up-turned curvature.

2.2.2 Stars and planets

During totality, one literally experiences night in the middle of the day. And because of the darkened sky, the brighter planets and stars can often be seen. Photographers can use the aid of planetarium apps or programs to make sure that they are able to select the correct field of view in order to capture these specks of light during totality.

2.2.3 Crescent sun projections

This effect is the only one on this list that is not visible during totality, but during the partial phases. It can also be seen during annular eclipses and partial eclipses. It is still a very interesting effect and is due to the projection of the partially eclipsing Sun through pin-holes. The pin-hole can be either man-made, as in a small hole pierced through a sheet of cardboard, or naturally occurring, as in holes through foliage. The projected image can be disconcerting as the Sun appears normally bright with the naked eye, and the projected image of a crescent-shaped Sun is often associated with the Moon. This effect can best be captured by having the image projected onto a flat, darkened screen that would improve the image's contrast.

2.2.4 Shadow bands

One of the most elusive effects are bands of shadow that race across the ground just before 2nd contact, and just after 3rd contact. These ripples of shadow are explained as being analogous to the ripples of light and shadow at the bottom of a swimming pool on a bright day. The atmosphere behaves as a fluid, as does the swimming pool's water, and refracts the light from the eclipsing Sun. The narrowness of the light from the Sun makes the refraction noticeable. This effect is best observed by laying out a flat white sheet on the ground, so that the contrast of the bands of shadow can be seen.

3 Capture Parameters

3.1 Location in Sky

Eclipses occur at varying times of the day, and therefore varying locations in the sky. The data for an eclipse can be obtained from sites like those from NASA[1]. Picking a site on the track of an eclipse will then provide valuable information including the site's latitude and longitude, the duration of totality, the date and times in UTC, and the altitude and azimuth of the various phases of the eclipse. For an eclipse lasting about 2 hours, the Earth will have rotated about 30 degrees, so a tracking mount would be helpful, but not mandatory, to counteract this movement.

With the Altitude and Azimuth of the 1st and 4th contacts, photographic and videographic equipment can be oriented to make sure that fields of view are appropriate, including watching out for obstacles or other problems. There are apps for phones available that will assist in this setup by using augmented reality to project the path of the eclipse onto the sky.

3.2 Chronology of Events

As the duration of a total solar eclipse is so short, understanding the sequence of events, many of which may be happening simultaneously, is critical to success. Mapping the events over the course of the eclipse is very informative as to how to capture the eclipse. Table 1 depicts when each of the effects discussed occur as the eclipse progresses. The effects marked with diamonds in their row occur at or around 2nd and 3rd contact and are the most difficult to capture because of their brevity. The effects that occur during totality are next in line as the time to capture them are also relatively short. Effects during the partial phases can be done at leisure as these 2 phases last about an hour each. Even then, the progress of the eclipse

Table 1. Chronology of Events

1st Contact — 2nd Contact — 3rd Contact — 4th Contact

	Partial Phase	Totality	Partial Phase
Duration	~1 hour	0-6.5 minutes	~1 hour
Corona		▒	
Diamond Ring		◇ ◇	
Bailey's Beads		◇ ◇	
Chromosphere		▒	
Prominences		▒	
Face of Moon		▒	
Sky Colour, Shape of Colour	▒	▒	▒
Stars, Planets		▒	
Crescent Sun Projections	▒		▒
Shadow Bands		◇ ◇	

marches on incessantly, so if images every 5 minutes were desired, these phases can still be pretty busy for the avid eclipse capturer.

3.3 Field of View and Speed of Capture

The various effects span a very wide range of field of views. Sky views are the widest, and wide-angle lenses and even fish-eye lenses can be used to capture these. Prominences would probably be on the other end of the scale, as they require super-telephoto lenses.

Similarly, the duration of each effect, and the time needed to capture the effect, varies just as much. During the partial phases, solar filters that allow only a very small percentage of light through, must be used. Even then, capturing the eclipsing Sun is relatively easy as the light from the Sun is abundant, even when it is almost totally occluded. Exposure times can be very short. However, during totality, solar filters should be removed as the amount of light falls dramatically. Exposure times vary significantly depending on what is being captured. Bailey's beads are fleeting, and needs to be captured with very short exposures and a high frame rate to ensure the right moment is caught. Capturing the dim outer edges of the corona requires long exposures of up to 30 seconds (10 seconds if not using a tracking mount).

The range of fields of view, and recommended camera focal lengths, for regular 35mm cameras at ISO 100, are provided in Table 2. As can be seen, some effects are best captured using video instead of photographs.

3.4 Equipment Considerations

Consolidating all the various effects to be captured, a range of equipment can be utilised.

3.4.1 Tracking Mounts, Tracking Heads, Tripods

Astronomical mounts including equatorial and Alt-Az mounts with their 1.5 or 2 inch tripods can be used effectively to track and capture the eclipse. However, these are sometimes not practical as they tend to be heavier and bulkier, and locations may be remote with limitations on luggage. Tracking heads are much more portable, but they may struggle with some of the equipment loads they need to support. A compromise is always needed for eclipses to provide the right combination. Aluminium tripods can be paired with the tracking heads, but for

Table 2. Recommended parameters

	Duration	Field of View	Recommended Eff Focal Length	Recommended Speeds
1st to 4th Contact	~2 hours	Zoom	~1000mm	Video
Corona	0-6.5 min	Zoom	200-500mm	1/8000 – 30 sec
Diamond Ring	<30 sec	Zoom	~1000mm	1/2000 sec
Bailey's Beads	<1 sec	Zoom	>2000mm	1/8000 sec
Chromosphere	0-6.5 min	Zoom	~1000mm	1/2000 sec
Prominences	0-6.5 min	Zoom	>2000mm	1/2000 sec
Face of Moon	0-6.5 min	Zoom	~1000mm	10 - 30 sec
Sky Colour, Shape of Colour	~2 hours	Wide	12-35mm	1/60 sec
Stars, Planets	0-6.5 min	Wide	18-35mm	~5 sec
Crescent Sun Projections	0-6.5 min	Normal	50-100mm	1/60 sec
Shadow Bands	~10 sec	Normal	35-50mm	Video

superior rigidity and portability, carbon fiber tripods are the way to go, at the expense of a heftier price tag.

3.4.2 DSLR cameras, CCD cameras, Lenses, Telescopes

The lens/sensor combination is extremely wide, with many brands and models that can be used. Even smartphone cameras can be used by many to capture the event, although they will never be able to match the size of lenses and sensor sensitivities of the larger equipment. Most DSLRs have a bracketing function, and this is very useful to capture events like Bailey's Beads and the Corona. CCDs would best be paired with a laptop (which can be a burden and an added point of failure) with software that can automate the capture process. Telescopes can be paired with CCDs or DSLRs, with afocal, prime or projection configurations all possible depending on the equipment available. Most simple refracting telescopes are usable, beginning from smaller ones with apertures of 50mm, with apochromatic telescopes providing better pictures due to the reduced chromatic aberrations.

When using DSLR cameras, it is best to capture photographs in raw mode, especially for the chromosphere, corona, and prominences. The dynamic ranges of these objects are very high so capturing in raw ensures that there is at least the ability to push the limits of processing to coax out the smallest of details.

3.4.3 360 Cameras

These cameras are excellent for capturing wide scenes, including effects on the ground (reaction of people, sounds, behaviour of fauna) as well as in the sky (colour of sky, changes in colours, 2nd Contact through 3rd Contact). It is best to locate these cameras at about head height (so that it captures views that appear to be from human height) and be mounted on monopods.

3.4.4 Drones

Drones are a relatively recent technological advancement. They provide incredible views from above and are amazingly intuitive to control. There are many legal restrictions with drones so understanding the region's regulations are important before using them. There are also many safety and ethical considerations for their use, including the fact that drones may obstruct, interfere or just ruin other observers' eclipse experiences. However, the ability to capture a terrestrial view from up high provides a stunning way to capture the eclipse effects on the environment, so navigating some of these obstacles are certainly worthwhile.

3.4.5 Webcams and Smartphones cameras

Smartphones are currently one of the more ubiquitous devices. Their photographic capability improves significantly year over year, and they even boast having 3 lenses with different fields of view in a single device. They are excellent for providing a whole variety "B-roll" footage of eclipses, and can also perform more complex tasks like time-lapse, hyper lapse, and panoramic images. The newer webcams also feature some of the newer sensors and are also getting very capable. Live streaming is an excellent, and also a relatively recent way of sharing the eclipse experience. However, an internet connection would be required, so this depends on whether the location would have this available.

3.4.6 Action Cameras

The small action cameras (most notably GoPro-type cameras) provide an excellent way to document eclipses. As they typically have wide angle lenses to capture more of the action, they are more suited to environmental captures. There are some newer configurations where zoom lenses can be fitted to these cameras to allow capturing Sun/Moon effects, but they are still very new and in its infancy as of the time of this writing.

3.5 Technological Aids

As technology progresses by leaps and bounds, the gadgets and applications that it brings are a real boon for eclipse capturing. Here are just some of the few items that are incredibly valuable to have onsite for capturing an eclipse.

3.5.1 Internet access

With roaming data, or even local data plans widely available in most developed or developing countries, the internet is now accessible at an observation site as long as it is not too remote. The internet is crucial for getting updated cloud cover satellite views and forecasts. If it were stable and fast enough, live streaming of the progress of eclipse becomes feasible. Realtime weather data is also important as there can be safety ramifications if storm warnings, fires, or other similar information is not available.

3.5.2 Smartphones and Smartphone Applications

Today's smartphones contain practically all the devices that had to be individually carried around a decade ago. GPS coordinates can now be obtained on the smartphone. Mapping software can show one's exact location, especially when inclement weather requires observers to "chase" the eclipse. Most if not all smartphones include a magnetometer, which can perform the role of a compass to find the correct azimuth. They would also have gravitational sensors which can function as an inclinometer to determine the altitude of the different contact positions in the sky.

A Bluetooth earbud is something that is undervalued during an eclipse, but is immensely useful especially when there is a lot going on. A simple recording can be played back to cue either a countdown to the different contacts of the eclipse, or to provide a reminder of what action is needed. For example, at 30 seconds before 2nd contact, a reminder to "remove solar filters" can be played. An earbud (and not just audio from the smartphone's speaker) is preferred as the environment will most likely be raucous, especially when it is shared with other eclipse observers, and screams of bewilderment and excitement will drown out even the loudest of our smartphone's speakers.

Apps are significant enough that they deserve their own mention. Planetarium programs such as Sky Safari or Stellarium, are incredibly useful to simulate the eclipse beforehand, as well as to determine eclipse views when on the move (for "chasing the eclipse" scenarios). There are also eclipse apps (e.g. EclipseDroid) that provides invaluable information during eclipses. These include eclipse timings, the countdowns to contacts, simulations of eclipse views, and even augmented reality, which uses the smartphone's camera to capture the environs, and project an overlay of the eclipse's contact positions onto the actual positions in the sky. This absolutely simplifies the setting up of the fields of view of camera equipment.

Finally eclipse apps, or camera-specific apps, can go so far as to control cameras and other devices. Apps like EclipseDroid provides scripting abilities, so that a sequence of shots (e.g. to capture the corona) can be prepared in advance, and triggered when necessary. There are also intervalometer apps that can control many of the established cameras, which would be useful for time-lapse or other automated imaging capabilities.

3.5.3 Laptops

Laptops can be all powerful, and many have brought them to capture eclipses and control more complex equipment. However, they are power hungry, and prone to many idiosyncrasies that may make them less than reliable, especially when compared with their simpler smartphones. The challenge is always whether one can afford the hiccup during those critical few minutes, so weighing its pros and cons is going to be required.

3.5.4 Power banks and spare batteries

Without power, many, if not all of the devices would not be able to function, and many of their built-in batteries would also not last an entire eclipse from 1st to 4th contact. Thanks to advances in technology, the density of batteries, namely the lithium ion variety, is incredibly high. Power banks are relatively inexpensive and very capable at providing portable power

Table 3. Recommended Equipment

	Tracking Mounts	DSLR/CCD + Telescope or Zoom Lens	360 Cameras	Drones	Action Cams, Smartphones
1st to 4th Contact	X	X	X	X	X
Corona	X	X			
Diamond Ring		X			
Bailey's Beads		X			
Chromosphere		X			
Prominences		X			
Face of Moon	X	X			
Sky Colour, Shape of Colour			X	X	X
Stars, Planets			X		
Crescent Sun Projections					X
Shadow Bands			X	X	X

in remote locations. Sealed Lead Acid batteries are also very suitable for portable power needs, but they mainly provide for devices that require 12V, vs the 5V that power banks provide.

4 Orchestrating the Capture

Considering that the average duration of totality of most eclipses is about 2 to 3 minutes, and that taking a photo could take as long as 30 seconds (not even considering the time required to review and check that photographs were exposed correctly and necessary adjustments made), the importance of exposing images and capturing them with clockwork precision becomes very apparent. Also, as there are different devices that are suited for different eclipse effects, it is best to deploy several, and/or multiple devices to optimise the capture of effects throughout the eclipse.

Table 3 lists a possible assortment of the equipment that can be used for capturing the various events. As can be seen, a DSLR or CCD sensor coupled with a zoom lens or telescope provides the narrow field of view needed to capture many of the more esoteric and technically difficult aspects of the eclipse. This setup in turn would not do a good job in capturing the environmental aspects which require a wider angle of view, and simpler, more portable devices should be used instead.

Fig. 4 depicts the physical deployment of one such setup to capture the eclipse. Starting from the centre, where the observers are, are 2 narrow field devices (DSLR or CCD sensor coupled with a zoom lens or telescope), one for capturing photographs, and the other for capturing video footage. There are 2 wide field cameras for video, one in front taking in footage of the observers and the background, and another behind the observers to capture the view from the rear. A 360-camera is placed somewhere central to the observing area, and preferably with an unobstructed view all around as this footage can be used for creating VR videos.

The last element depicted in Fig. 4 is a drone in the air. The height of the drone depends on the view it is trying to capture – lower down to capture more details in the observers' reactions, or higher up to capture the changing colours in the landscape and environs. Other very important considerations for drones would require a whole separate discussion.

If there were several technically proficient operators of each device, orchestrating the capture would be a much simpler task. But more often than not, due to cost reasons, only one or two people would be operating all the devices. It is still possible for a single person to manage all the devices, provided he or she is well-versed in the various technologies. A very

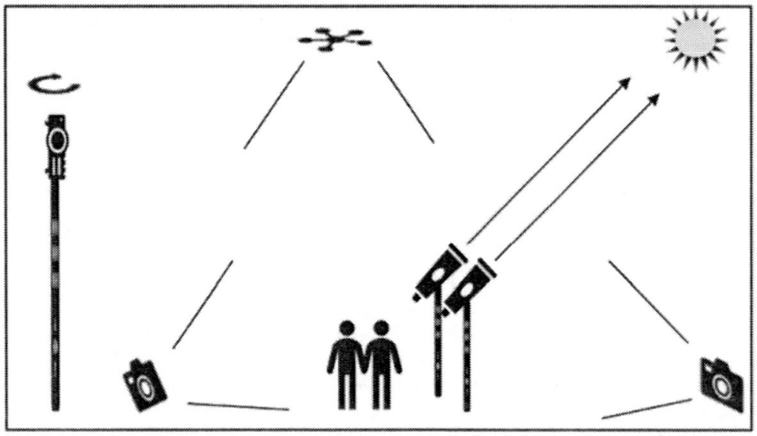

Fig. 4. Physical Deployment of Equipment

focused and organised mind, with good diagnostic skills (as there will always be last-minute issues), are paramount in pulling this off.

Fig. 5 shows the timeline of activities that needs to be coordinated, with the majority of activities occurring about 10 minutes prior to 2nd contact, and 5 minutes after 3rd contact. The assumptions in this timeline are that the front and rear cameras will be continuously filming the entire eclipse from 1st to 4th contact, and that there will also be photographs of the progressing eclipse (which can be replaced with videos if the camera is on a tracking mount) in 5-minute intervals. Practicing the sequence of activities prior to the day, preferably with a simulation of the eclipse on a tablet or smartphone, will go a long way in making sure that all the kinks are ironed out before eclipse day. Of note in the activities listed, is the "Look at the Eclipse" reminder, since the human eye's amazing dynamic range can take in the view incredibly fast, and store that in memory, so as to recreate the experience later in post production.

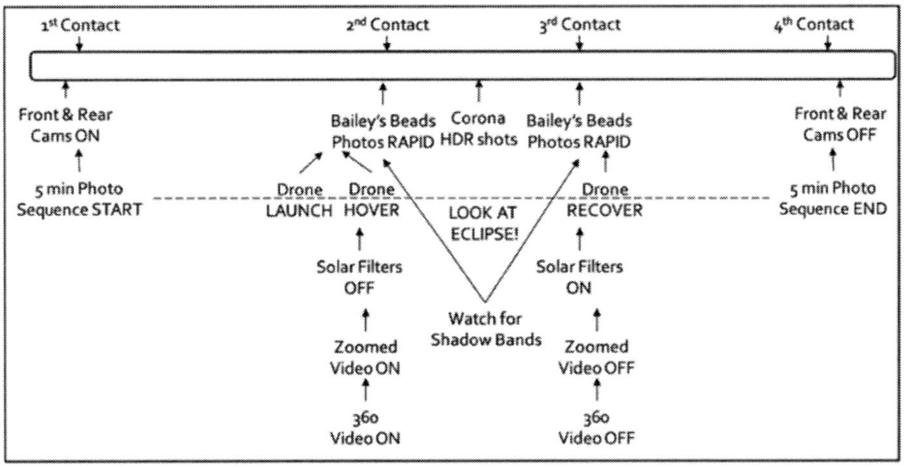

Fig. 5. Actions and Event

5 Post processing and production

After a successful eclipse, there is an immense satisfaction and relief as a significant amount of preparation, planning and work would have gone into capturing the fleeting event. However, the work after the eclipse, to process the photographs as well as to assemble a video to share the experience, is probably just as daunting a task for many.

5.1 Adobe Photoshop – Photographs

Photoshop is the gold standard for digital photography processing. Basic processing of jpg and raw files will not be discussed as there are many other sources for this information, so this should be able to create very good images (assuming of course that they were captured well) of the chromosphere, prominences, the diamond ring, and Bailey's beads. Here are some of the more unique forms of output for photos taken during an eclipse:

5.1.1 Eclipse GIFs

GIFs are a form of animation, and is assembled by combining several frames of the eclipsing Sun. This is best done with a zoomed in view of the Sun, and each photo should be processed such that the Sun is always in the same size and orientation (i.e. de-rotated if taken via a non-equatorial mount). It is best to use Photoshop's guides to create 4 reference lines (top and bottom extremities and left and right edges) from an image of the Sun before 1st contact or after 4th contact.

All images that are to be used to create the GIF should be loaded into their own layer, and in the order in which they are to be animated – typically top to bottom representing the animation from 1st to 4th contact. All layers should be hidden, and the Sun on the topmost layer should be carefully centred and sized. If the Sun before 1st contact is not available, as with locations where the eclipse has already started before it can be seen, then a reference after 4th contact can be used.

The reference image can be toggled on and off, and its opacity modified, in order to size and align each and every image in the stack of images. This can be painstaking, but manual alignment is normally superior to the automated programs, as they aren't able to differentiate and align eclipse images which generally have very few detectable alignment features. It is important to note that it is not recommended to use the Moon as a reference for sizing and orientation. This is because the Moon, at least for total eclipses, is larger than the Sun, and it is highly likely that the Moon will be overexposed in most frames, and therefore its edges are going to be blown out and imprecise. Also, having the Sun stationary as the reference object, and having the Moon move across its face, has a more natural effect.

After all the images have been assembled, sized and aligned, the GIF can be assembled using Photoshop's "Timeline" feature. This allows all the frames to be assembled in its correct order. The timing of each frame can be manipulated to achieve various effects, and a suitable timing would be 0.5 seconds for each frame from 1st to 2nd contact, and 3rd to 4th contact, and to extend the timing during totality between 2nd and 3rd contact to 1 to 2 seconds or more. The GIF can also be set to infinitely loop so that it will replay itself after its sequence.

The GIF output file can then be created using "File | Save for Web", where the image type can be set to GIF. Other settings recommended would be to use the "Adaptive" colour map with 256 colours, and dither pattern of "Noise".

5.1.2 The Solar Corona

Recreating the solar corona that we see with the naked eye in a photograph is one of the more satisfying tasks. It is basically creating a high dynamic range image from about 8 to 10 different images with exposures ranging from 1/8000th of a second to 30 seconds. The raw versions of these images should be used and processed using Adobe Camera Raw. Recommended settings are as in Table 4.

Once each image is optimised, they should be saved in 16-bit colour, and then loaded into Photoshop in individual layers and converted into smart objects. They can then be combined using the Mean function (Layer | Smart Objects | Stack Mode | Mean) and filtered (Filter | Camera Raw Filter) by adjusting for clarity, shadows and highlights. The layers can be aligned visually by nudging each layer one at a time. The Moon should be in the exact centre and set to a level that makes it all black, with any prominences showing as pink highlights.

Finally, the layers can be merged together, and then duplicated. One layer can be filtered with a radial blur filter, and using image calculations, subtract the blurred image from the original image to form an alpha channel image that can be used to sharpen the corona streamers by setting its mode to "overlay". This technique can be used more than once if desired for more sharpening.

Table 4. Adobe Camera Raw settings

Parameter	Setting
Shadows	+83
Clarity	+85
Sharpen	70
Radius	1.0
Detail	25
Masking	Sharpen till noise begins to appear, then decrease below that.
Luminance	33
Detail	50
Color	26 (or just to remove fringing)
Detail	50
Lens Correction	Remove Chromatic Aberration

5.2 Videos.

There is a significant amount of creativity in putting together a video, and many variations can be found on the internet. The main concepts and ideas behind the video are discussed here more than the actual use of the tool. Adobe Premiere is probably one of the better tools that can be used to achieve any effect desired, though its learning curve is commensurately higher than other simpler tools.

5.2.1 Story boarding

It is important to decide what the video should convey to its audience, whether it is to be a personal story, a travelogue, or an educational video. Story boards are a great way to chart out the flow visually. Some common themes in these story boards include the eclipse track on the globe, the position of observation on the track, the geographical journey, the eclipse surroundings, eclipse timing and location in the sky, and of course of the actual eclipse itself.

5.2.2 Time considerations

As the partial phases take an hour or more in actual time, audiences would definitely be bored to watch this in real time. Having these phases compressed would create a better flow for the video, and a compression rate of 30 to 100 times would probably work well. Conversely, totality itself, and especially at the instances of 2nd and 3rd contacts, the actual time is very short, and goes by too fast for viewers. It is worth extending these times by either slowing them down, repeating them several times, or even replaying the same scene from different angles.

5.2.3 Multiple source considerations

With multiple video sources recording at the same time, some of the better video editing tools have the capability to synchronize the different sources together so that it is simple to switch from one view to another without losing continuity in the resulting video. In order to do this, traditional movie producers would use a clapper board. The frame where the clapper claps and the sound that is made, is used to synchronize both audio and video streams. In the case of the devices used to capture the eclipse, the best moment to "clap" is not before 1st contact, which is what may seem to be the obvious moment. The best time to do this is actually when the last device before 2nd contact is turned on because there are many devices with limited power (e.g. the drone), so it is best to have all devices active before the synchronization event is played. This should then be able to be used to synchronize all the streams produced.

Audio as mentioned is also an important stream, and fortunately, most recording devices include an audio track as well. Only one of the many tracks though, should be used as a master track, and this should ideally be the longest (i.e. has been recording throughout) as well as the clearest. Advanced video editing software, like Adobe Premiere, do have the capability to use each source's audio track to perform the synchronization function with. The sounds recorded during an eclipse are often unique and spectacular, as the reaction of the crowd and the surroundings during totality are often missed during the eclipse, and only realised when the recording is replayed.

The last aspect to mention is colour grading, which is the blending of the different video source's individual colour palettes. Most of the advanced video editors have this function, and to enable this, it is often best to have a white sheet of cardstock displayed or visible to all the video sources, so that colour grading can be done easily. One of the more important settings for video devices is their white balance setting. Most devices will have automatic white balance, or AWB for short, to automatically adjust the white balance point. This would actually be detrimental for eclipse capturing, as the changes in colour is actually desirable. AWB should therefore be turned off for video devices, but calibrated correctly using the reference white cardstock.

6 Summary

This article describes many of the amazing spectacles that can be observed during a total eclipse. Some of the tools and techniques that can be used to capture a total eclipse were discussed, and these are always being refined as technology, knowledge and experience improves with each eclipse. For anyone familiar with project management and the skills involved, they should be able to recognize that these skills are exactly what is best brought to bear to orchestrate the capture of an eclipse. That, and a good grasp of technology, to maximize its use as well as the ability to rapidly diagnose and solve any issues that emerge will ensure a successful capture. It is certainly best to experience a total solar eclipse in person, but for those unwilling or unable to travel, there is no better way to experience it than through the work of others, either in their photographs or in their videos. Hopefully, the information here would contribute in some way to the success of these astronomical hunters.

References

[1] NASAweb: Solar Eclipse Page: Solar Eclipses Past and Future, https://eclipse.gsfc.nasa.gov/solar.html (accessed 15 Nov 2019)

High-resolution solar imaging for citizen science

Alfred Yong Liang Tan[1,2,*]

[1]Elected Fellow, Institute of Physics, Singapore
[2]National Junior College, 37 Hillcrest Road, Singapore

Abstract. High-resolution solar imaging, essential for research into the Sun is usually limited to high-end expensive telescopes found in professional observatories using adaptive optics technology. However, a Schmidt-Cassegrain Telescope (SCT) or a refractor telescope equipped with a narrow-band energy rejection filter and a suitable solar spectrum filter, employing appropriate data acquisition and post-processing techniques can provide an amateur astronomer a cost-effective means to high-resolution solar imaging.

1 Introduction

In the last decade, students in Singapore schools have produced solar images in Hydrogen-Alpha, Calcium-K and White-light wavelengths. There was a keen interest to produce high-resolution solar images of the Sun for research work [1].

Students in several schools in Singapore conducted extensive tests from 2016 to 2019 to customise different solar telescope configurations and build a cost-effective observatory for producing high-resolution solar images [2, 3]. National Junior College used the findings of the trials and built a High-Resolution Solar High Observatory (HISRO). Following its commissioning in July 2019, students from NJC and other participating schools have used HISRO for their research work (See Figure 1).

2 Equipment for High-resolution Imaging

Different locally-customised configurations of high-resolution imaging rigs were tested. A typical configuration for high-resolution imaging is shown in Figure 2. The setup comprises:

- Dielectric-Energy Rejection Filter
- Host telescope
- Solar Spectrum Filter
- Astro camera with barlow or reducer
- German Equatorial Mount
- Field Imaging Instrumentation Console

2.1 Dielectric-Energy Rejection Filter (D-ERF) [4]

* Email : yltansg@gmail.com

This project used D-ERFs with diameter ranging from 165mm to 356mm to enable the host telescope to retain its full-aperture resolving power while reducing the thermal energy going through the telescope to the solar spectrum filter.

Fig. 1. HISRO was opened by Dr. C. Alex Young, NASA Goddard. 500 students from 12 schools attended the opening and solar observation on that day

Fig. 2. A typical setup for high-resolution solar imaging

The D-ERF reduced the heat stress within the telescope and also reduced the thermally-induced seeing effects as well. A single-band H-Alpha 165mm D-ERF was used on a 150mm refractor telescope. A 295mm D-ERF (for 279mm SCT) and 214mm D-ERF (for 203mm SCT) were used for tri-band (H-Alpha, Calcium-k and G-Band) imaging. A 365mm D-ERF was used for H-Alpha imaging with a 356mm SCT. Typical H-Alpha tranmission and Calcium-K transmission were about 90% and 60% respectively (See Figure 3). "Leakage" into the unwanted wavlength was typically under 5%.

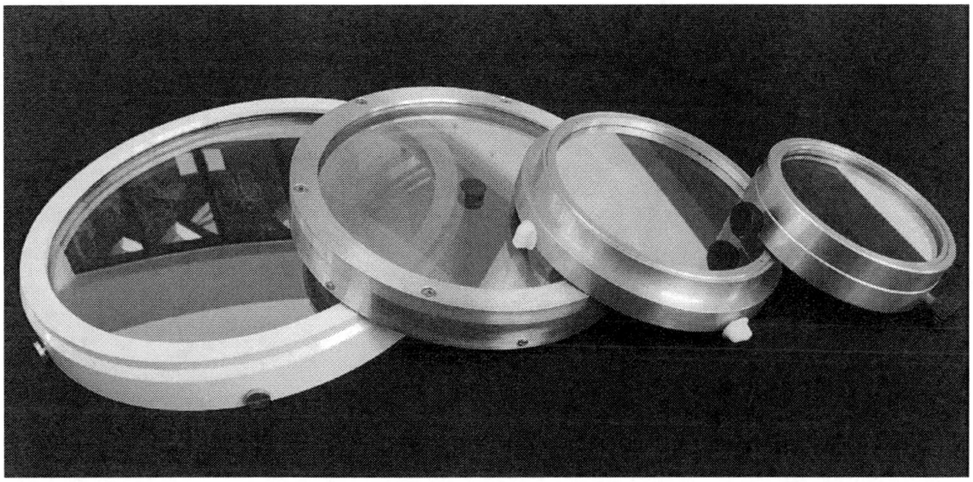

Fig. 3. Four types of dielectric-Energy rejection filters (D-ERF) ranging from 165mm to 356mm were used on different-sized refractors and SCTs

2.2 Host Telescope [5, 6]

Typical SCTs, with SCT aperture ranging from 203mm to 356mm and a 150mm refractor were used as host telescopes.

2.3 Solar Spectrum Filter [5, 6]

The project teams tested a range of commerically available etalons and integrated them into donor telescopes fitted with full-aperture D-ERF to form different customised solar telescope configurations. Favourite components included a solid mica etalon and air-spaced etalons without central obstruction for both H-Alpha and Calcium-k imaging. Other components included G-Band filter, H-Alpha blocking and Calcium-K module (See Figure 4).

2.4 Astro camera with barlow or reducer.

An astro CCD camera with a high acquisition speed was used for high-resolution solar imaging. To achieve the required imaging scale the necessary barlow or reducer was added before the astro camera.

2.5 German Equatorial Mount.

To provide smooth tracking of the Sun, a typical German Equatorial Mount with polar alignment was used to carry the solar telescope and imaging train.

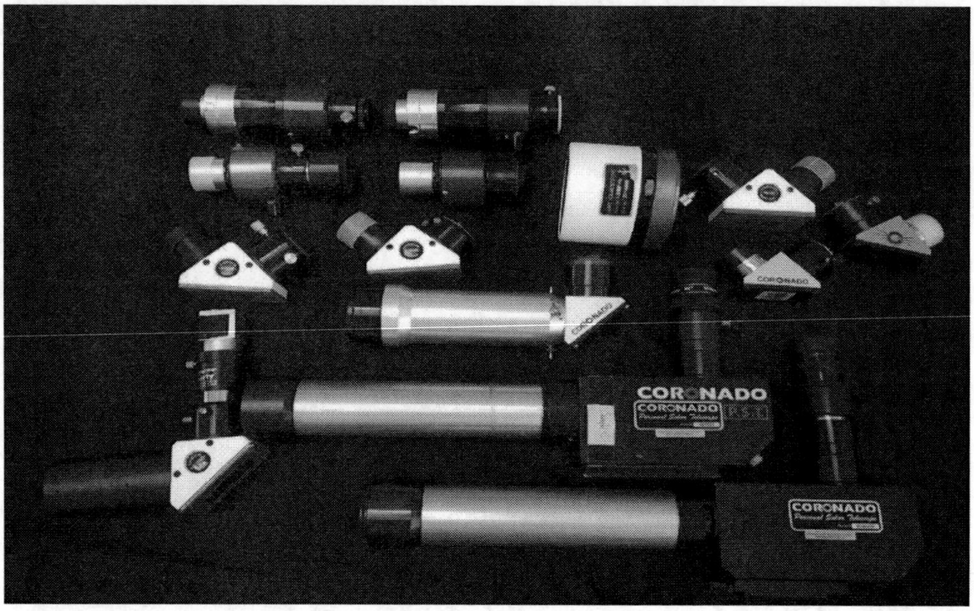

Fig. 4. Different types of etalons were used to provide the desired bandpass for high-resolution imaging

2.6 Field Imaging Instrumentation Console.

To have a dark workspace in the field, the team constructed a field imaging instrumentation console. The enclosure shielded the recording equipment from sunlight and enabled easy video acquisition in the field [7].

3. Open field observatory [8]

NJC High-resolution Solar Research Observatory has four imaging mounts holding three SCTs and one refractor telescope for high-resolution solar imaging. It was built with the following considerations:

- **Open concept.** It was located at a site for easy access for students to conduct solar observation and research work. (See Figure 5)
- **Imaging environment.** Seeing conditions affect the quality of imaging. Heated air from the ground will produce turbulence and affect the quality of the imaging. The team selected a site surrounded by greenery and situated on a relatively high ground.
- **Setup time.** Students attend lessons during daytime and they need to commence imaging as soon as possible when the seeing conditions permit high-resolution imaging. HISRO was able to commence imaging within 5 minutes of reaching the observatory.

The team achieved the above-mentioned requirements by building HISRO at a location where students will walk past in between lectures. The telescopes were left permanently on the mounts and protected from the elements using a newly devised 3-layer covering method (See Figure 6). Students are able to commence imaging after a 5-minute setup time. They have to remove the protective covers, deploy the field imaging instrumentation console and power up the telescopes before commencing the imaging session.

Fig. 5. Aerial view of National Junior College High-Resolution Solar Research Observatory

Fig. 6. Telescopes were left on the mount when not in use. A 3-layer covering method, devised by Mr A. Tan, provided protection of the equipment from the elements. This method of covering and protection has been trialed for a year. All the equipment did not show signs of deterioration in Singapore's humid and wet weather.

4 Workflow

Atmospheric turbulence randomly distorted the wavefronts emanating from the Sun and was the primary obstacle in achieving diffraction-limited resolution in high-resolution solar imaging. The following workflow was employed to reduce the adverse effect of atmospheric turbulence.

4.1 Imaging window [9]

The best quality images were acquired in the imaging window from 8:15 am to 10:30 am local time, throughout the year. This early window of opportunity will avoid jet stream emanating from the heated ground and rooftop. Nevertheless, the observatory was able to produce images of decent quality during the other times of the day due to the general good seeing condition in Singapore.

4.2 Data acquisition [10]

The camera has to be set up for each telescope configuration for optimum performance. To obtain individual captured frames with the best signal/noise ratio, the researcher needs to set the camera at the fastest shutter speeds with the least gain (brightness amplification) for least noise in the final image. Exposure was typically set below 5 msec to freeze the seeing. The capture rate was usually maintained close to 100 fps to prevent a long duration capture leading to changes in the solar features. Gamma setting was set to provide good image contrast to aid the focussing of the image. During capture, gamma moderation was necessary to ensure that subtle solar details within dark regions were not lost. Typically each video capture will last about 30 seconds of about 2000 captured frames.

4.3 Lucky imaging [11, 12]

Before image restoration techniques were applied, the frames captured in the most favourable atmospheric conditions (lucky frames) were selected using solar image processing software. During good seeing, up to 1,000 best frames were selected and stacked by the same solar image processing software to form a single resulting image with a high signal-to-noise ratio. There was a need to experiment and try stacking different number of selected frames to find the best resulting image.

4.4 Post-processing [9]

As part of image restoration, the final stacked image was further processed using Lucy Richardson deconvolution and unsharp masking techniques to produce a sharp and yet not overly sharpened high-resolution image revealing the fine features in the captured region of the Sun. The post-processed image could be further enhanced with an image editing software.

5 Results

5.1 M5.0 flare.

A hybrid solar telescope using a typical 150mm achromatic refractor fitted with a full-

aperture D-ERF achieved an image scale of 0.45 arcsec per pixel (See Figure 7). It captured a M5.0 flare on 23 July 2016. (See Figure 8) The A&G News & Reviews in Astronomy & Geophysics Volume 58, Issue 6 December 2017 featured these images as a contribution to citizen science [13]. (See animated sequence at https://tinyurl.com/M5-0-flare).

Fig. 7. 150mm H-Alpha solar telescope

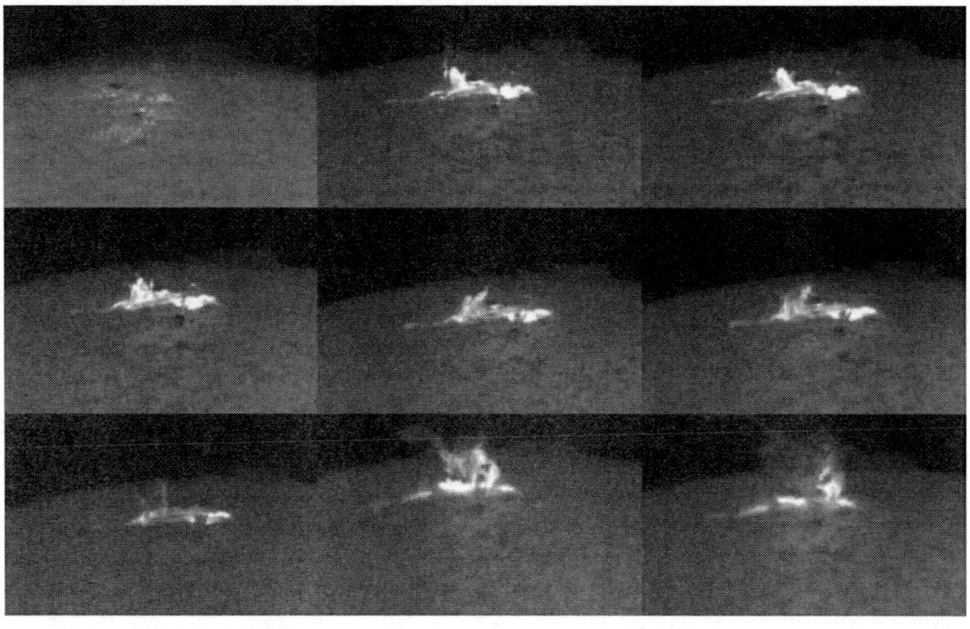

Fig. 8. Sprays and surges with the solar chromospheric network and fibril structures

5.2 G-Band granulations.

Figure 9 shows a 430nm G-Band image of granules at the top of the convection zone in a typical quiet photospheric region. Granule down to 1000 km can be observed. It was taken on 26 April 2017 with a 203mm SCT fitted with a tri-band (H-Alpha,

Calcium-K and G-band) full-aperture D-ERF, a commercial G-Band (filter passband 10 nm) and x5 barlow. (See Figure 10) With good seeing, the imaging rig, with an image scale of 0.23 arc sec/pixel reached the diffraction limit for this wavelength and this telescope size.

Many amateur astronomers were amazed by what a typical 203mm SCT could achieve. They celebrated this achievement and compared it favourably with a 1.6m specialized solar telescope equipped with advanced adaptive optical system at the National Solar Observatory. This breakthrough sparked off a lively discussion amongst amateur astronomers [14].

Figure 11 shows a G-Band image of twin sunspots in a photospheric region. This image was taken later in the day after the Figure 9 image. The seeing condition had deteriorated and thin clouds veiled the Sun. The gain on the camera was stepped up and a more noisy image resulted. These 2 images (i.e. Figure 9 versus Figure 11) showed the vital part seeing condition played in the eventual quality of image. Subsequent trials showed that a bigger image scale should be used to improve the image quality when seeing has gone downhill.

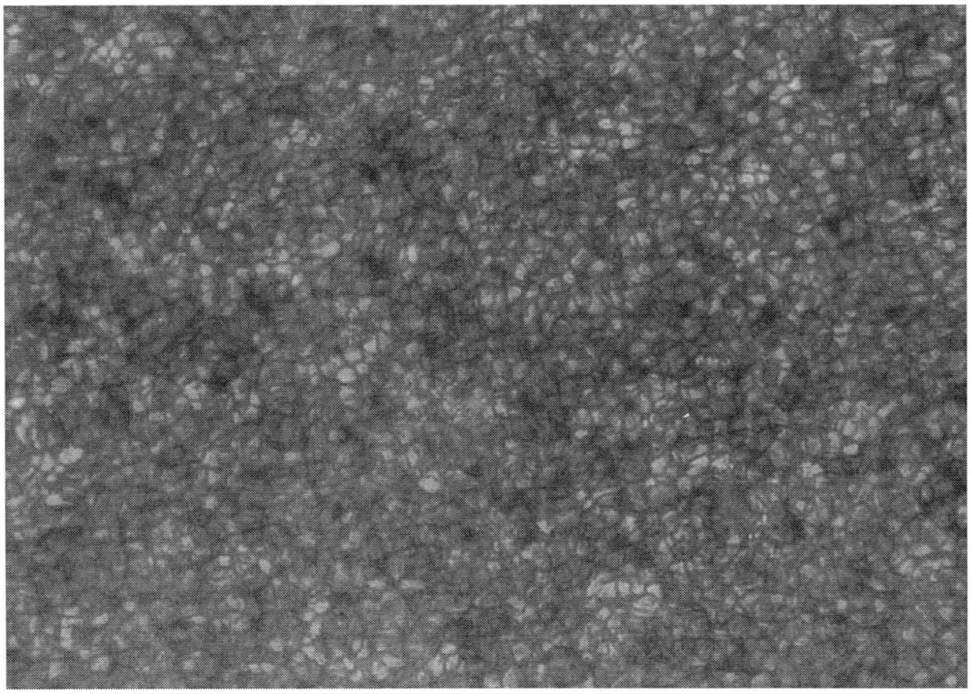

Fig. 9. A typical quiet photospheric region imaged by a 203mm SCT in the G-band. Image scale was 0.23 arc sec per pixel.

Fig. 10. G-Band imaging with a 203mm SCT in Paya Lebar Methodist Girls' School

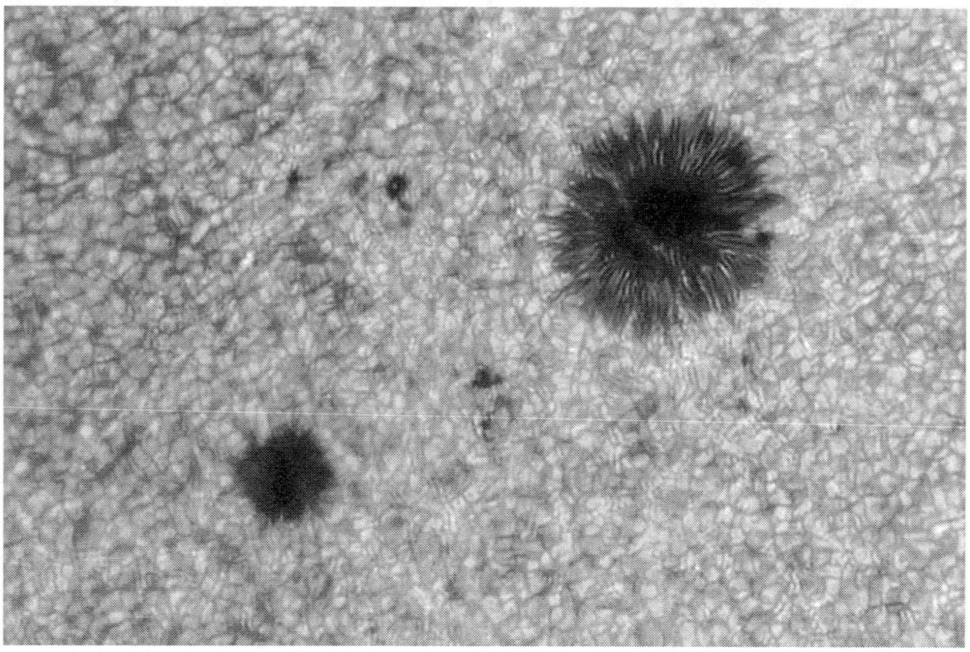

Fig. 11. G-band image of Sunspots in the photospheric regions. Image scale was 0.23 arcsec per pixel

5.3 Survey of the upper chromosphere during solar minima.

A 279mm SCT fitted with a tri-band (H-Alpha, Calcium-K and G-band) full-aperture D-ERF, a solid mica H-Alpha etalon and x0.8 reducer were used to survey the Sun's upper chromosphere in H-Alpha wavelength from January 2020 and February 2020 (See Figure 12).

Different sized reducers (e.g. x 0.5 to x0.8 reducers) with different length spacers were used to achieve different image scale ranging from 0.15 to 0.2 arcsec per pixel. Higher image scales were used in poorer seeing conditions.

Some of the collected images in Figures 13, 14 and 15 revealed fine fibril details, magnetic field topology and the dynamic nature of a quiet solar chromosphere [15]. As can be observed in Figure 14 and the evolution of fibrils could be traced. A boiling chromosphere was observed in an animated gif from a 5 minute recording session [16].

Fig. 12. A 279mm SCT was used to survey the chromosphere with a H-Alpha mica etalon

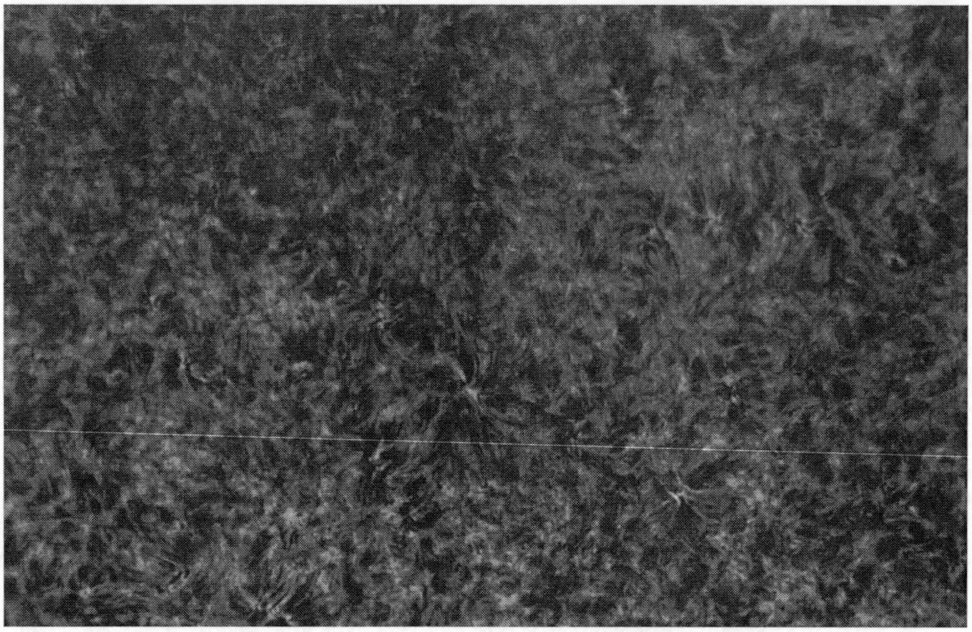

Fig. 13. Filamentary structures portray the magnetic field topology in the chromosphere

Fig. 14. A 5-minute animated gif showed the dynamic nature of a quiet solar chromosphere [16]

Fig. 15. Bright spots were frequently seen near multi layers of plasma fibrils

5.4 Survey of the lower solar chromosphere

The same 279mm SCT fitted with a tri-band (H-Alpha, Calcium-K and G-band) full-aperture D-ERF was fitted with a calcium-K wedge for surveying the lower solar chromosphere. (See Figure 16). With a x2 barlow, an image scale of 0.22arcsec per pixel was achieved.

The chromospheric network as observed with this configuration are shown in Figure 17 and Figure 18 [17-19].

Fig. 16. A 279mm SCT fitted with a Calcium-K wedge was used to survey the lower solar chromosphere

Fig. 17. Magnetic bright spots observed tend to concentrate in a granular pattern

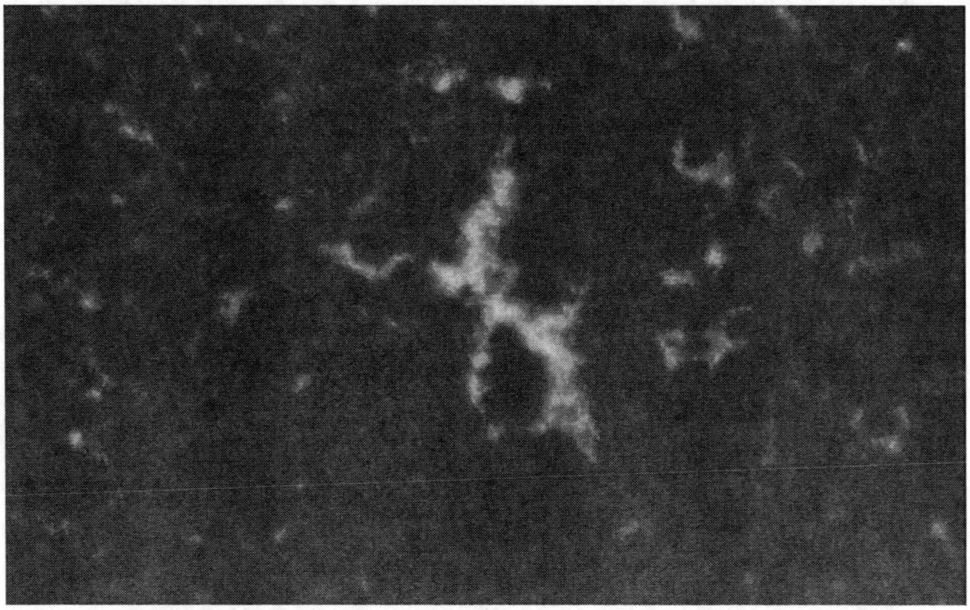

Fig. 18. The chromosphere network as observed with the 279mm SCT. Magnetic bright points formed a granular pattern.

5.5 Fibrils

The same 279mm SCT shown in Figure 1 was fitted with a tri-band (H-Alpha, Calcium-K and G-band) full-aperture D-ERF and an air-spaced etalon without central obstruction to image the dark elongated fibril structures in H-Alpha wavelength [20].

Dark structures were observed in the penumbra of a sunspot, radically expanding around the umbra and creating filamentary structure in the active region (See Figure 19).

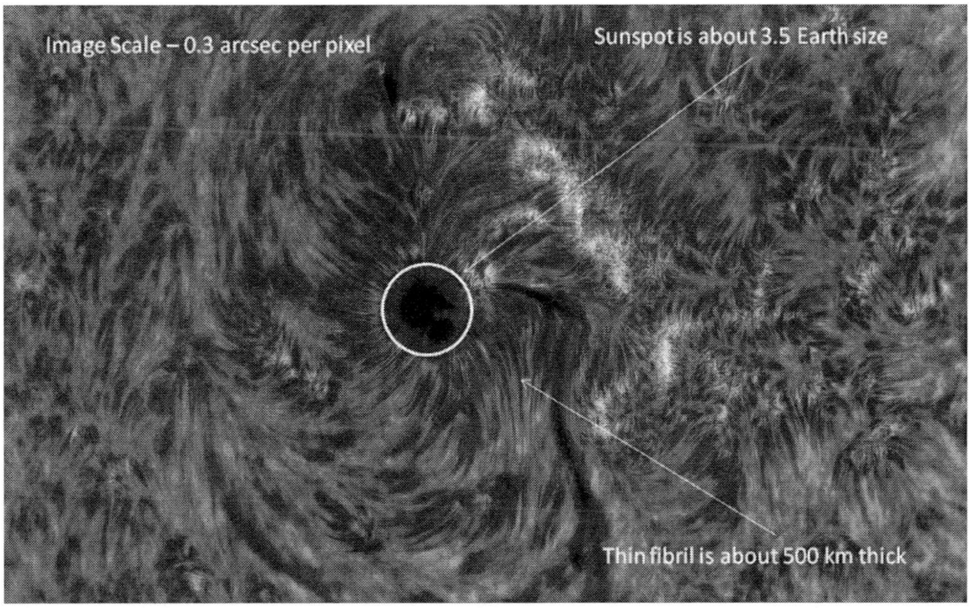

Fig. 19. Dark elongated fibril structures surround the sunspot

6 Conclusion

The different solar telescopes, when used with appropriate data acquisition, post-processing and image restoration workflow, showed great potential as cost-effective solutions for amateurs to do high-resolution solar research. More work was needed to overcome atmospheric seeing when imaging with a 356mm SCT and improving image restoration work processes to produce better quality images [21].

Acknowledgments

This material was a consolidation of the investigative work to establish different configurations of optimized high-resolution solar telescopes for different imaging requirements and to build an open solar observatory for students' research. Students and teachers from Paya Lebar Methodist Girls' School, National Junior College and other partner schools participated in this effort. They used the prototype telescopes for their respective teaching and learning activities. The images that they have produced provided valuable feedback for the project. I acknowledge the expert advice provided by fellow astronomers from https://solarchatforum.com/. I also acknowledge the support given by my wife, Mei for the purchase of all the equipment, encouraging me to press on since I began the project in 2016 and proof reading this material.

References

1. Research activities for amateurs, http://www.astrosurf.com/luxorion/research.htm, (2020)

2. G. Garcia, ALPO Feature: The Sun High Resolution Solar Imaging, *The Strolling Astronomer*, **45**, No.3, 16 -20, (2003)

3. H. Wang, C. Liu, *Signatures of magnetic flux ropes in the low solar atmosphere observed in high resolution,* Frontiers in Astronomy and Space Sciences, (2019)

4. Description of Airylab 235mm DERF, http://astrograph.net/Airylab-235mm-D-ERF-Filter-including-Custom-Cell, (2020)

5. https://airylab.com/hat-solar-telescopes, (2020)

6. https://www.skyatnightmagazine.com/reviews/telescopes/airylab-203mm-hydrogen-alpha-chromosphere-telescope/, (2020)

7. https://petapixel.com/2015/05/19/a-6-ikea-storage-box-makes-for-a-thrifty-collapsible-laptop-sun-shade/, (2020)

8. https://www.nytimes.com/2002/07/18/garden/design-notebook-in-backyards-starry-appeal.html, (2020)

9. https://solarnutcase.livejournal.com/18572.html, (2020)

10. K. M. Harrison, Camera exposure times, *Imaging Sunlight Using a Digital Spectroheliograph* (2016) Springer International Publishing, pp. 123

11. Law, N.M, Mackay, C.D.,Baldwin, J.E., *Lucky imaging: High angular resolution imaging in the visible from the ground,* (2006)

12. Beckers, J.M., *Improving solar image quality by image selection*. (1989)

13. L. Fletcher, A. Berlicki, A. K. Awasthi and D. Gronkiewicz, *Recruiting flare hunters for citizen science,* The A&G News & Reviews in Astronomy & Geophysics, **58**, Issue 6 December 2017

14. https://solarchatforum.com/viewtopic.php?f=4&t=22409, (2017)

15. G.Tsiropoula, K. Tziotziou, I. Kontogiannis, M.S. Madjarska, J.G. Doyle, Y. Suematsu, , *Solar fine-scale structure*. pp. 8-10, (2012)

16. https://1.bp.blogspot.com/-K8-cR4Irj9M/XigpQYpdG5I/AAAAAAAApEE/Dmh-RdrHXcYDJWXrj56rtell-hhX8_dMACLcBGAsYHQ/s1600/10frames.gif

17. J. Leenaarts, M. Carlsson, L.R. Van Der Voort, *On fibrils and field lines: The nature of Hα fibrils in the solar chromosphere,* (2018)

18. V. H.Hansteen, M. Carlsoon, *The solar atmosphere*, pp 13-16, (2019)

19. J.L. Jenkins, *Observing the Sun, Lower chromosphere, K-Grains, Chromospheric network*, pp. 50 - 52, (2013)

20. https://www.seeker.com/fibril-flowers-on-suns-surface-photo-1769556565.html (2015)

21. A. Macrobert, Sky&Telescope, How to successfully beat atmospheric seeing, https://skyandtelescope.org/astronomy-equipment/beating-the-seeing/ (2006)

Modified Statistical Analysis of SNe1a Data

Lisa Goh[1,*] and *Cindy* Ng[1,**]

[1]Department of Physics, National University of Singapore, 2 Science Drive 3 Singapore 117542

Abstract. We review an improved maximum likelihood analysis of the Type 1a Supernova (SNe1a) data. We calculate the profile likelihood in the Ω_m-Ω_Λ parameter space by conducting a parameter sweep across the 8 SNe1a parameters, using a Markov Chain Monte Carlo (MCMC) optimization algorithm. This improved analysis, which does not assume arbitrary values for the uncertainties, has the advantage of being bias-free as compared to the original analysis. We use the Joint Lightcurve Analysis (JLA) dataset containing 740 SN1a data samples for our study, and compare among 5 different models: the ΛCDM model, the flat wCDM model, its non-flat generalization, as well as two dynamical w(z) parametrizations. We find that the ΛCDM model is favoured over the other models, and the best fit values based on this model are Ω_m=0.40 and Ω_Λ=0.55. Interestingly, in most of the contour plots we obtain, the line of no acceleration is crossed at $2\sim3\sigma$ confidence levels, which is similar to the results published by Nielsen et al, the original authors who introduced the improved maximum likelihood analysis. When we generalize the wCDM model to the dynamical w(z) parametrizations, the evidence for cosmic acceleration becomes even weaker. This raises the question of how secure we can be of an accelerating expansion of the universe.

1 Introduction

In 2011, Perlmutter, Schmidt and Riess were awarded the Nobel Prize in Physics for their discovery of the accelerating expansion of the Universe [1][2], by studying distant Type 1a Supernovae (SNe1a). They noticed that these supernovae were dimmer than expected of a constantly expanding Universe, and after some data analysis, concluded that the cosmological term Λ in Einstein's field equations had to be non-zero, or there existed an additional dark energy term.

Eight years on, surveys such as the Sloan Digital Sky Survey [3], the Supernova Legacy Survey [4], the Hubble Space Telescope [5], ESSENCE [6], LOSS [7], Harvard-Smithsonian Centre for Astrophysics CfA1-4 survey [8], and the Carnegie Supernova Project [9] have collected large amounts of SNe1a data in the past two decades covering a wide range of redshifts from $0.01 < z < 0.1$. In the last one year alone, there has also been new surveys conducted, such as the PAN-STARRS 1[10] and Dark Energy Survey[11], which has brought the total number of supernova samples to an unprecedented number of over 1000 samples. With an increasing number of datasets, we can place further constraints on

*e-mail: e0008130@u.nus.edu
**e-mail: cindy.ng@nus.edu.sg

the exact quantities of dark energy and matter in the Universe, consequently allowing us to improve or debunk our current theories and models.

However, data will only be useful if we can analyse them accurately in order to obtain meaningful results, and draw correct conclusions. Many different statistical data analysis methods have been proposed as means to analyse SNe1a data, based on both Bayesian [12][13] and frequentist [14] approaches. Particularly in a paper that we will be looking at in more detail, the methods they have used led them to the conclusion that a constant expansion cannot be ruled out. Their results have been verified by some [13][15], which is interesting to note. Therefore this paper aims to study their method more in-depth and put forth some modifications to it, by exploiting the advantages of both Bayesian and frequentist statistics. We will then use our new method on an expanded range of cosmological models to test their goodness of fit, and perhaps give additional insight towards the probability of a constant rate of expansion.

This paper has been arranged as such: Section 2 introduces the cosmological theories behind Universe expansion, Section 3 will explain and justify our proposed methodology, in Section 4 we present our results, Section 5 will provide a discussion of our results, and finally we give a conclusion in Section 6.

2 Theory

The main source of data being used to investigate the expansion of the Universe are Type 1a supernovae, also known as SNe1a.

SNe1a are characterised by their lack of hydrogen lines, unlike Type II Supernovae. Since all SNe1a occur at the same mass through the same mechanisms, their peak luminosities are roughly constant across all SNe1a, thus making them good standard candles with which to measure distances in the Universe. Using the equations

$$\mu = m - M \tag{2.1a}$$

$$\mu = 25 + 5\log(d_L) \tag{2.1b}$$

where m is the apparent magnitude and M is the absolute magnitude (≈ -19),the distance modulus μ and luminosity distance d_L can be known.

The graph of luminosity against time (known as the lightcurve) of every SNe1a show very similar shapes, therefore researchers make use of the lightcurve data in order to obtain distance estimates of μ.

2.1 Models of the Universe

The models of the Universe which we will be testing our data against are the concordance model, the ΛCDM model, and other variations of it, namely the wCDM model and its parametrizations.

2.1.1 ΛCDM Model

The current concordance model of the Universe, developed as a response to the discovery of accelerating expansion, is known as the ΛCDM, or Λ Cold Dark Matter model. Here,

Λ represents dark energy. It has a negative pressure $p = -\rho c^2$ which, when substituted into the (ii) components of the momentum-energy tensor $T_{\mu\nu}$, (where i =1,2 and 3), gives an accelerating expansion. Therefore it is now widely accepted to exist in our Universe, although the exact way in which it manifests itself is still shrouded in mystery.

Based on this model, the expression for the evolution of H as a function of redshift z, known as the Hubble function, is given by the relation

$$H(z)^2 = H_0^2[\Omega_m(1+z)^3 + \Omega_k(1+z)^2 + \Omega_\Lambda] \tag{2.2}$$

where $E(z) = \sqrt{\Omega_m(1+z)^3 + \Omega_k(1+z)^2 + \Omega_\Lambda}$, and Ω_m, Ω_Λ and Ω_k are the mass density, dark energy density and curvature density parameter respectively.

2.1.2 wCDM Model

An extension of the ΛCDM model is the wCDM model, where we are concerned with the EoS parameter of dark energy, w. In the ΛCDM model, it is assumed to take the value of -1. We model this evolution with 2 different parametrizations, and test them against our data. Most wCDM models assume a flat Universe with $\Omega_k = 0$, however we shall generalize this in order to find the best fit value for Ω_k.

The first parametrization (hereafter referred to as parametrization A) is to treat w as a scalar value unchanging with redshift. The equation for $H(z)$ can thus be written as

$$H(z)^2 = H_0^2[\Omega_m(1+z)^3 + \Omega_k(1+z)^2 + \Omega_\Lambda(1+z)^{3(1+w)}] \tag{2.3}$$

Parametrization B defines the evolution of w as a linear function of redshift [16]:

$$w(z) = w_0 + w_1 z \tag{2.4a}$$
$$H(z)^2 = H_0^2[\Omega_m(1+z)^3 + \Omega_k(1+z)^2 + \Omega_\Lambda(1+z)^{3(1+w_0-w_1)}e^{3w_1 z}] \tag{2.4b}$$

Parametrization C defines the evolution of w as a first order Maclaurin expansion of redshift [16], and is also known as the Chevallier-Porlarski-Linder parametrization [17]:

$$w(z) = w_0 + w_1 \frac{z}{1+z} \tag{2.5a}$$
$$H(z)^2 = H_0^2[\Omega_m(1+z)^3 + \Omega_k(1+z)^2 + \Omega_\Lambda(1+z)^{3(1+w_0+w_1)}e^{3w_1[1/(1+z)-1]}] \tag{2.5b}$$

We find the best fit values for w, w_0 and w_1 for each parametrization.

We compare the data to the model by calculating μ as given by equations (2.1a) and (2.1b). The equation for d_L, the luminosity distance, is given by

$$d_L = \begin{cases} (1+z)\frac{d_H}{\sqrt{\Omega_k}} \sinh\left(\sqrt{\Omega_k}\int_0^{z_0} \frac{dz}{E(z)}\right), & \text{if } \Omega_k > 0 \\ (1+z)d_H \int_0^{z_0} \frac{dz}{E(z)}, & \text{if } \Omega_k = 0 \\ (1+z)\frac{d_H}{\sqrt{|\Omega_k|}} \sin\left(\sqrt{|\Omega_k|}\int_0^{z_0} \frac{dz}{E(z)}\right), & \text{if } \Omega_k < 0 \end{cases} \tag{2.6}$$

where $d_H = c/H_0$ and $E(z) = H(z)/H_0$.

3 Methodology

3.1 SALT2 Lightcurve Fitting

We will be using the modified equation for calculating μ, based on Guy et al. [18] to address the fact that the brighter the SNe1a, the wider its lightcurve, and the bluer it will appear to be. Therefore

$$\mu_B = m_B^* - M + \alpha \times x_1 - \beta \times c \tag{3.1}$$

where m_B^* is the new notation for the apparent magnitude, and the values of α and β are constant throughout the dataset being studied. Each SNe1a will be stretched along the time axis and colour corrected by a certain amount to fit the template model. This method has been adopted by many survey teams when presenting lightcurves of new SN1e data, and is the method we will be adopting to analyze our SNe1a dataset.

The dataset that we will analyze is the Joint Lightcurve Analysis (JLA) dataset published by Betoule et al [14] (hereafter B14). It combines data from the Sloan Digital Sky Survey (SDSS) and the Supernova Legacy Survey (SNLS) to give 740 SNe1a data samples.

They present their errors in the form of a 3×740 by 3×740 covariance matrix for the parameters m_B^*, x_1 and c. The total covariance matrix $\mathbf{C}$ is a sum of 8 covariance matrices that address two main sources of uncertainties:the systematic uncertainty and statistical uncertainty. It is therefore given by

$$\mathbf{C} = \mathrm{C_{cal}} + \mathrm{C_{model}} + \mathrm{C_{host}} + \mathrm{C_{bias}} + \mathrm{C_{pecvel}} + \mathrm{C_{nonIa}} + \mathrm{C_{dust}} + \mathrm{C_{stat}} \tag{3.2}$$

where $\mathrm{C_{cal}}$ is uncertainty matrix associated with the calibration, $\mathrm{C_{model}}$ is the light-curve model uncertainty, $\mathrm{C_{host}}$ is the mass step uncertainty, $\mathrm{C_{bias}}$ is the bias correction uncertainty, $\mathrm{C_{pecvel}}$ is the peculiar velocity uncertainty, $\mathrm{C_{nonIa}}$ is the contamination due to non-type 1a SN, and $\mathrm{C_{dust}}$ is the uncertainty due to dust extinction in the Milky Way.

3.2 The Maximum Likelihood Estimation Method

The maximum likelihood estimation method was introduced by Nielsen et al.[12] (hereafter N16), which is based on calculating the likelihood ratio instead of the χ^2 value. Therefore the best fit value is the value that gives a maximum likelihood value, and not a minimum χ^2. Furthermore, they do not just treat Ω_m and Ω_Λ as the only parameters to find the best fit values of, but also do a sweep across the set of 8 parameters $\{M_0,x_{1_0},c_0,\sigma_{M_0},\sigma_{x_{1_0}},\sigma_{c_0},\alpha,\beta\}$. Here, we give an introduction of their method.

3.2.1 Bayesian Statistics

In Bayesian statistics, the probability density is given by [19]

$$\mathrm{P}(A|B) \propto \mathrm{P}(B|A)\mathrm{P}(B) \tag{3.3}$$

where $\mathrm{P}(A)$ is the prior distribution, $\mathrm{P}(A|B)$ is the posterior distribution and the value of $\mathrm{P}(B|A)$ is the likelihood ratio. In this case, A is the model and B is the data. Therefore we are calculating the probability of the model being a good fit, given the data.

The method used in N16 is based on an extension of Bayesian statistics, known as the Hierarchical Bayesian Model. This model should be used, instead of regular Bayesian statistics, when there are multiple parameters with prior distributions that depend on other parameters [20]. This is indeed true for the data we have, and we shall see that the set of parameters $\{M_0, x_{1_0}, c_0, \sigma_{M_0}, \sigma_{x_{1_0}}, \sigma_{c_0}, \alpha, \beta\}$ depend on one another. In this case, the Hierarchical Bayesian probability density is given by

$$\mathrm{P}(\gamma, \alpha|\beta) \propto \mathrm{P}(\beta|\alpha, \gamma)\mathrm{P}(\alpha|\gamma)\mathrm{P}(\gamma) \tag{3.4}$$

Here, $\mathrm{P}(\gamma, \alpha|\beta)$ is the posterior distribution, $\mathrm{P}(\gamma)$ is the prior distribution and $\mathrm{P}(\beta|\alpha, \gamma)\mathrm{P}(\alpha|\gamma)$ is the new likelihood ratio. In the context of SNe1a data analysis, we shall see what α, β and γ represent, and why we need to include one extra parameter to draw more accurate conclusions based on the data.

3.2.2 Calculating Likelihood L

The values of m_B^*, x_1 and c obtained are not the true values due to various sources of noise and error, both systematic and statistical. We shall denote these observed values as $\hat{m}_B^*$, $\hat{x}_1$ and $\hat{c}$, and the true values without a hat. Therefore, adopting the form of (3.4), we can express the total probability density as

$$P[\theta, (M, x_1, c)|(\hat{m}_B^*, \hat{x}_1, \hat{c})] \propto P[(\hat{m}_B^*, \hat{x}_1, \hat{c})|(M, x_1, c), \theta]P[(M, x_1, c)|\theta]P(\theta) \tag{3.5}$$

where the likelihood ratio is $L = P[(\hat{m}_B^*, \hat{x}_1, \hat{c})|(M, x_1, c), \theta]P[(M, x_1, c)|\theta]$, and θ denotes the model. In terms of (3.4), θ is γ, and the data we obtain, $(\hat{m}_B^*, \hat{x}_1, \hat{c})$, is β. We also wish to infer the true parameters from the observed parameters, therefore (M, x_1, c) is α, where M is the absolute magnitude which is not taken to be a constant value for every SNe1a here. The likelihood can be written in terms of an integral

$$L = \int P[(\hat{m}_B^*, \hat{x}_1, \hat{c})|(M, x_1, c), \theta]P[(M, x_1, c)|\theta]dMdx_1dc \tag{3.6}$$

The first factor of the integrand addresses the discrepancy between the obtained values and the true values, and the second factor addresses the difference between the true values and the model. We shall now derive an equation to calculate, and subsequently maximize, the value of L.

When we plot a histogram of $\hat{x}_1$ and $\hat{c}$, we see that they roughly follow a Gaussian distribution.

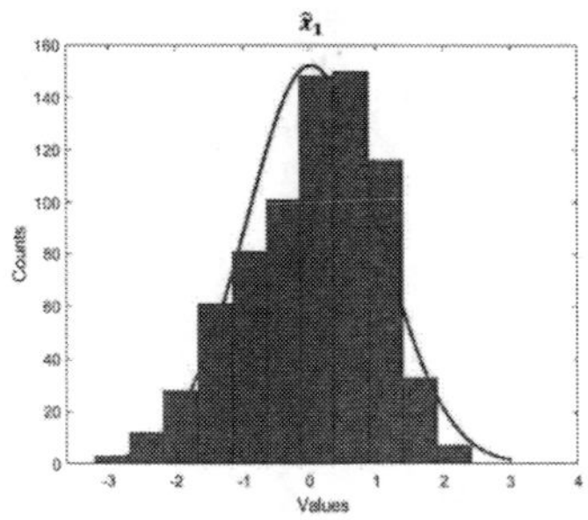

(a)

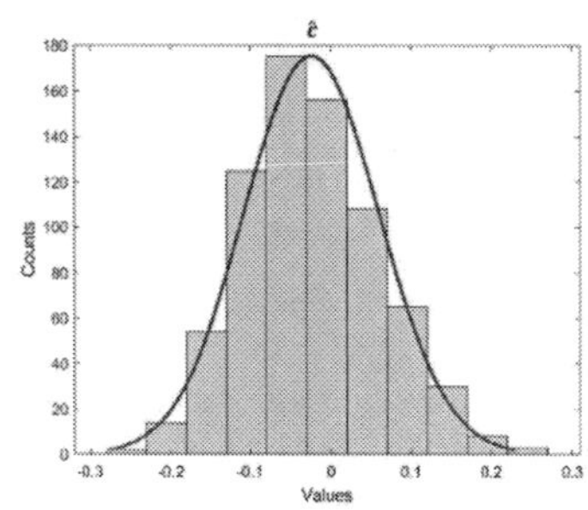

(b)

Figure 1: Histograms of $\hat{x}_1$ and $\hat{c}$, with a Gaussian distribution overlaid. We can see that they roughly follow a Gaussian distribution, therefore we assume that the true values also follow a Gaussian distribution.

Assuming a Gaussian distribution of M as well, we model the probability distribution of M, x_1 and c as a Gaussian function:

$$P(M) = (2\pi\sigma_{M_0}^2)^{-1/2}\exp\left(-[(M - M_0)/\sigma_{M_0}]^2/2\right) \tag{3.7a}$$

$$P(x_1) = (2\pi\sigma_{x_{1_0}}^2)^{-1/2}\exp\left(-[(x_1 - x_{1_0})/\sigma_{x_{1_0}}]^2/2\right) \tag{3.7b}$$

$$P(c) = (2\pi\sigma_{c_0}^2)^{-1/2}\exp\left(-[(c - c_0)/\sigma_c]^2/2\right) \tag{3.7c}$$

where M_0, x_{1_0} and c_0 are their mean values and σ_{M_0}, $\sigma_{x_{1_0}}$ and σ_{c_0} are their variances. The total probability density is a product of these three terms: $P(M, x_1, c|\theta) = P(M)P(x_1)P(c)$. For N SNe1a data points, we combine the three parameters to form a vector $Y = \{M_1, x_{1_1}, c_1, ..., M_N, x_{1_N}, c_N\}$. The total probability density is then

$$P(Y|\theta) = |2\pi\Sigma_l|^{-1/2}\exp\left(-(Y - Y_0)\Sigma_l^{-1}(Y - Y_0)^{\mathrm{T}}/2\right) \tag{3.8}$$

where $Y_0 = \{M_0, x_{1_0}, c_0...\}$ is a $3N$ vector of the mean values, and $\Sigma_l = \mathrm{diag}(\sigma_{M_0}^2, \sigma_{x_{1_0}}^2, \sigma_{c_0}^2)$ is the $3N \times 3N$ diagonal matrix of the variances. The absolute symbol denotes the determinant of the matrix. (3.8) corresponds to the second term that we will substitute into (3.6) in order to calculate L.

Now, we deduce the equation for the first term in (3.6). We define the $3N$ vector of observed values $\hat{X} = \{\hat{m}_{B1}^*, \hat{x}_{11}, \hat{c_1}..., \hat{m}_{BN}^*, \hat{x}_{1N}, \hat{c_N}\}$ and the $3N$ vector of true values $X = \{m_{B1}^*, x_{11}, c_1..., m_{BN}^*, x_{1N}, c_N\}$. The probability density of the observed data X given some true parameters $\hat{X}$, $P(\hat{X}|X, \theta)$, is

$$P(\hat{X}|X, \theta) = |2\pi\Sigma_d|^{-1/2}\exp\left(-(\hat{X} - X)\Sigma_d^{-1}(\hat{X} - X)^{\mathrm{T}}/2\right) \tag{3.9}$$

where Σ_d is the $3N \times 3N$ covariance matrix which is referred to as **C** in B14. However, N16 did not include the mass step correction C_{host}, as it does not make a significant difference to the results, which was also demonstrated by [13]. To convert m_B^* to M, we define the $3N$ vector $\hat{Z} = \{\hat{m}_{B1}^* - \mu_1, x_{11}, c_1, ..., \hat{m}_{BN}^* - \mu_N, x_{1N}, c_N\}$, where μ is the theoretical value of the distance modulus based on the model, and the $3N \times 3N$ block diagonal matrix $A = \{1, 0, 0; -\alpha, 1, 0; \beta, 0, 1; ...\}$, to get the relation $\hat{X} - X = (\hat{Z}A^{-1} - Y)A$. Then, $P(\hat{X}|X, \theta) = P(\hat{Z}|Y, \theta)$. This is the first term in (3.6). Combining everything,

$$\begin{aligned} L &= \int P(\hat{Z}|Y, \theta)P(Y|\theta)dY \\ &= |2\pi\Sigma_d|^{-1/2}|2\pi\Sigma_l|^{-1/2}\int dY \exp\left(-(Y - Y_0)\Sigma_l^{-1}(Y - Y_0)^{\mathrm{T}}/2\right)\exp\left(-(Y - \hat{Z}A^{-1})A\Sigma_d^{-1}A^{\mathrm{T}}(Y - \hat{Z}A^{-1})^{\mathrm{T}}/2\right) \\ &= \left|2\pi(\Sigma_d + A^{\mathrm{T}}\Sigma_l A)\right|^{-1/2}\exp\left[-(\hat{Z} - Y_0A)(\Sigma_d + A^{\mathrm{T}}\Sigma_l A)^{-1}(\hat{Z} - Y_0A)^{\mathrm{T}}/2\right] \end{aligned} \tag{3.10}$$

Just like in R98, we can construct contour plots in the $\Omega_m - \Omega_\Lambda$ parameter space, this time by finding the set of values for the 8 parameters $\{M_0, x_{1_0}, c_0, \sigma_{M_0}, \sigma_{x_{1_0}}, \sigma_{c_0}, \alpha, \beta\}$, that give a maximum likelihood for each pair of $(\Omega_m, \Omega_\Lambda)$. This is known as the profile likelihood L_P. Denoting ϕ as the 8 parameters, which we treat as the nuisance parameters as we are not plotting them in the parameter space, and θ as the interesting parameters Ω_Λ and Ω_m,

$$L_P = \max_\phi L(\theta, \phi) \tag{3.11}$$

The likelihood value L is related to χ^2 by the formula $L = \exp\left(-\chi^2/2\right)$—a maximum likelihood corresponds to a minimum χ^2 value. Therefore in terms of χ^2, (3.10) becomes

$$\chi^2 = 2\pi\left|(\Sigma_d + A^{\mathrm{T}}\Sigma_l A)\right| - (\hat{Z} - Y_0A)(\Sigma_d + A^{\mathrm{T}}\Sigma_l A)^{-1}(\hat{Z} - Y_0A)^{\mathrm{T}} \tag{3.12}$$

which is a much easier equation to work with. Therefore, we calculate the χ^2 values using (3.12) instead for every $(\Omega_m, \Omega_\Lambda)$, and plot the contour lines of 1σ, 2σ and 3σ.

3.2.3 Comparison to the Least χ^2 Method

The Maximum Likelihood Estimation (MLE) method does not only vary Ω_m and Ω_Λ, but also the other 8 parameters $\{M_0, x_{1_0}, c_0, \sigma_{M_0}, \sigma_{x_{1_0}}, \sigma_{c_0}, \alpha, \beta\}$. An advantage the MLE method has over the χ^2 method is that it does not add an arbitrary uncertainty σ_{μ_0} (see [1])), which N16 argues can be adjusted to create biased results towards a preferred model, in this case the ΛCDM model. Therefore the χ^2 method only tests for a goodness of fit to the assumed model, but does not check if it is indeed the correct model. On the other hand, the calculation of L in the MLE method solely depends on the data obtained without adding uncertainties whose value differs for each SNe1a data sample.

From the results obtained in N16, the contour plot significantly crosses the line of no acceleration, therefore they argue that a non-accelerating expansion cannot be ruled out. Other papers have referenced their method with modifications [21] or additional constraints [15] to come to the conclusion that current evidence still favour an accelerating expansion. Therefore it is interesting to expand this new method to explore other Universe models or datasets to see what results we might obtain from it, and to see if our Universe is indeed expanding at an accelerating rate, and by how much.

3.3 Markov Chain Monte Carlo Optimization

The method we will be using to find the best fit values of $\{M_0, x_{1_0}, c_0, \sigma_{M_0}, \sigma_{x_{1_0}}, \sigma_{c_0}, \alpha, \beta\}$ is the Markov Chain Monte Carlo (MCMC) method. This method, based on Bayesian statistics, is widely used in many computational fields to approximate the distribution of multiple parameters—here, we aim to obtain the distributions of all the 8 parameters and find the set of 8 values which gives the maximum likelihood value. We adapt an MCMC code by Pitkin [22] and write our own code to calculate the likelihood ratio and the theoretical value of μ based on the model. The likelihood ratio calculation is based on the MLE method outlined in Section 3.2.2.

Pitkin implements an affine invariant ensemble method based on Goodman and Weare [23], a more sophisticated version of the Metropolis-Hastings method that is faster and can handle highly non-symmetric distributions.

4 Results

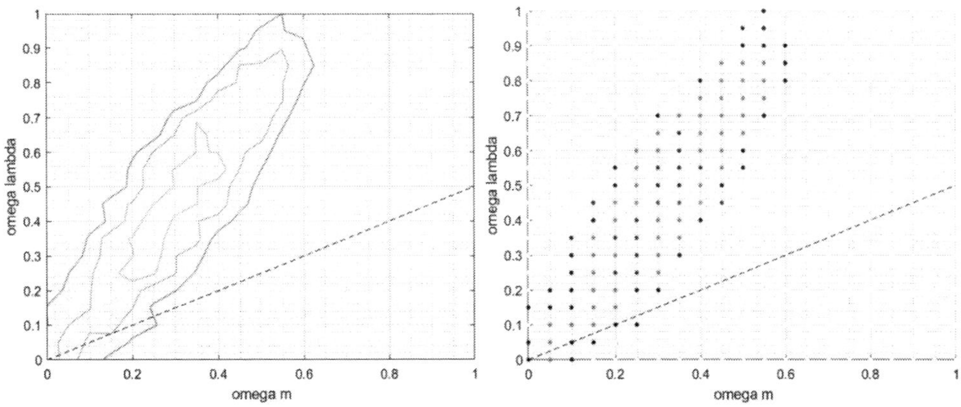

Figure 2: Contour plot (left) and scatter plot (right) of Ω_m against Ω_Λ for the ΛCDM model, using the MLE and MCMC method, with 1σ, 2σ and 3σ lines drawn. The dotted line represents the line of no acceleration. The best fit values of Ω_m and Ω_Λ are 0.40 and 0.55 respectively, with a minimum χ^2 value of -231.209.

For the ΛCDM model, we run through Ω_m and Ω_Λ from the values of 0 to 1, with a step size of 0.05. The plot we obtain is comparable to N16 where the 1σ and 2σ lines cross the line of no acceleration. Unlike N16, the contours are not smooth, which is due to the fact that an MCMC method was used— [13] also obtained ragged contours when they implemented their own MCMC algorithm.

Next, we test the flat wCDM model, following [15]. We do a parameter sweep across Ω_m and w for $\Omega_k = 0$, ie a flat case.

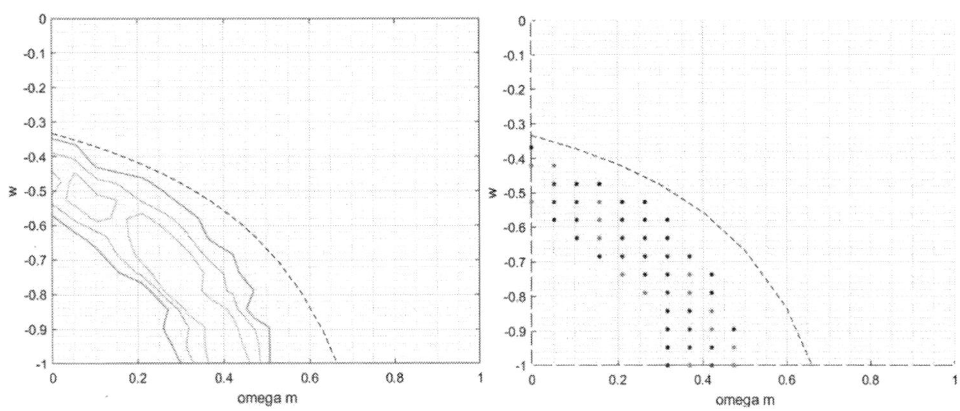

Figure 3: Contour plot (left) and scatter plot (right) of Ω_m against w for $\Omega_k = 0$ with 1σ, 2σ and 3σ lines. The dotted line is the line of no acceleration. We see that in our plot, the contour lines do not touch the line of no acceleration. The minimum χ^2 value is -229.6220, for $\Omega_m = 0.4211$ and $w = -1$.

Now, we generalize to a non-flat wCDM model where we do a parameter sweep across Ω_m, Ω_Λ and w.

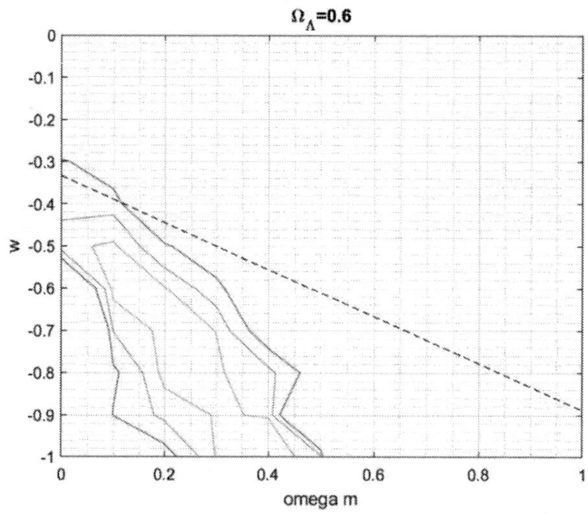

Figure 4: Plot of w against Ω_m for $\Omega_\Lambda = 0.6$, with 1σ, 2σ and 3σ contour lines corresponding to a difference in χ^2 values for 3 parameters, since we run through 3 parameters instead of just 2. The best fit value is $\Omega_m = 0.4$,$w = -1$ and $\Omega_\Lambda = 0.6$ with a χ^2 of -229.8145.

Following [16], we run through the values of $-2 < w_0, 0, -2 < w_1 < 8$ and Ω_m for a flat wCDM parametrization B model. We present plot of w_0 against Ω_m for interesting values of w_1.

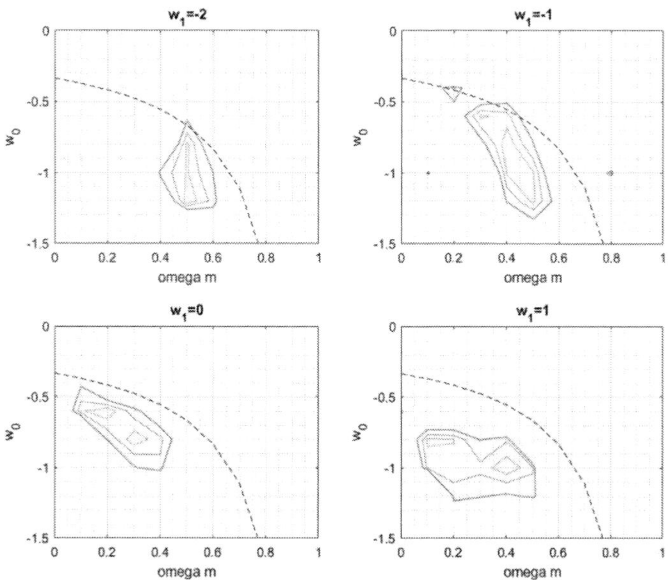

Figure 5: Plots of w_0 against Ω_m for values of $-2 < w_1 < 1$, with 1σ, 2σ and 3σ contours. The best fit values are $\Omega_m = 0.4$, $w_0 = -0.8$, $w_1 = -1$, with a χ^2 value of -229.4125.

We run through the same range of w_0 and w_1 as parametrization B for parametrization C and present our results, again with plots of Ω_m against w_0 for interesting values of w_1.

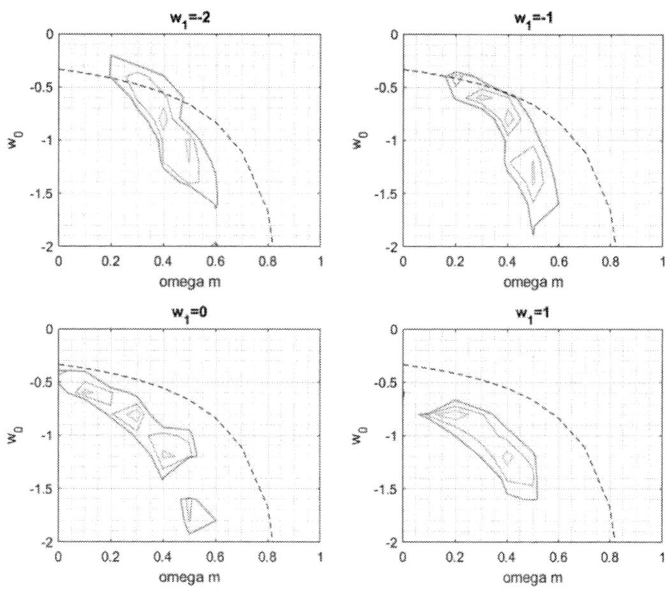

Figure 6: Plots of w_0 against Ω_m for values of $-2 < w_1 < 1$, with 1σ, 2σ and 3σ contours. The best fit values are $\Omega_m = 0.4$, $w_0 = -0.8$, $w_1 = -1$, with a χ^2 value of -227.4537.

5 Discussion

5.1 ΛCDM Model

Our results for the ΛCDM model differ slightly from N16 as the contour lines cross the line of no acceleration to a greater extent. Our minimum χ^2 value of -231.209 is also lower than the value published by N16 (-214.97). The table below shows how our best-fit Ω_m and Ω_Λ values compare to other works:

Table 1: Comparison of our best fit values of Ω_m and Ω_Λ against literature values

	Ω_m	Ω_Λ
Our work	0.40	0.55
N16	0.341	0.569
Sharriff et al	0.340	0.524
R98 (MLCS)	0.2551	0.7143

We see that our best-fit values are within 1σ of N16 and Sharriff et al's results, further lending clout to the accuracy of our results. This shows that our MCMC code is sufficiently accurate in finding the maximum likelihood and optimizing over the 8 parameter space.

5.2 wCDM Model

5.2.1 Parametrization A

We present the best fit values obtained for both the flat and non-flat wCDM models for parametrization A.

Table 2: Best fit values of Ω_m,Ω_Λ, Ω_k and w for flat and non-flat models of Parametrization A

Model	Ω_m	Ω_Λ	Ω_k	w
Flat	0.4211	0.5789	0	-1
Non-flat	0.4	0.6	0	-1

We see that the best fit values for the flat and non-flat models are in agreement with each other. Moreover, for both cases, the contour plot is much wider than that obtained by Haridasu et al, especially in the non-flat case, where the 3σ contour crosses the line of no acceleration. This differs from their claim that the results for an accelerating expansion is reinforced in the wCDM model, particularly when the constraint for a flat Universe is removed since the contours are wider than in the flat case. Therefore the claim for accelerating expansion should not be made purely based on assuming a flat Universe

5.2.2 Parametrizations B and C

We present the best fit values for Parametrizations B and C below:

In the best fit plots (ie. $w_1 = -1$) for both parametrizations, we see that the contour lines also cross the line of no acceleration to an even greater extent than in parametrization A ($\sim 2\sigma$ level). By generalizing the evolution of w, the evidence for constant acceleration becomes stronger. Perhaps if we extend these parametrizations to include the non-flat case, the data will become even more consistent with a constant acceleration model.

Table 3: Best fit values of Ω_m, w_0 and w_1 for Parametrizations B and C

Parametrization	Ω_m	w_0	w_1
B	0.4	−0.8	−1
C	0.4	−0.8	−1

5.3 Model Comparison

To assess which model gives the best fit of our data, we calculate the Akaike Information Criterion (AIC) and the Bayesian Information Criterion (BIC), two values that are often calculated to test the goodness of fit of different statistical models. The formulae to calculate AIC and BIC are given by [24] [25]:

$$\text{AIC} = 2k - 2\ln(L) \tag{5.1a}$$

$$\text{BIC} = \ln(n)k - 2\ln(L) \tag{5.1b}$$

where k is the number of parameters estimated, n is the number of data samples (in this case, the number of SNe1a data samples), and L is the maximum likelihood. The best fit model is the model with the lowest AIC and BIC values. However, the AIC and BIC test does not give an absolute benchmark on the goodness of fit of a model, only a comparison between models.

We find that the ΛCDM model has the lowest AIC and BIC values, and here, we present the ΔAIC and ΔBIC values for the other cosmological models as compared to the ΛCDM model: Based on the AIC and BIC values, the ΛCDM model is still the best fit

Table 4: ΔAIC and ΔBIC values for the five different models tested

Model	ΔAIC	ΔBIC
ΛCDM	0	0
Flat wCDM	1.5870	1.5870
Non-flat wCDM (Param. A)	3.3945	8.0012
Flat wCDM (Param. B)	3.7965	8.4032
Flat wCDM (Param. C)	5.7553	16.9686

cosmological model given the data, followed by the flat wCDM model. Moreover, the results from their generalizations serve to further validate their best fit value results, and the contour plots obtained using the MLE method provide interesting insights regarding the extent to which an accelerating expansion is certain. In almost all of our results, an accelerating expansion is only consistent at $\sim 2\sigma$ confidence levels. As such, constant expansion of the Universe cannot be ruled out, especially in the case of generalized cosmological models.

6 Conclusion

In this paper, we have used a modified statistical method to analyze data from the JLA SNe1a data set against 5 different cosmological models. This statistical method is based on a Maximum Likelihood Estimation and an MCMC optimization procedure to obtain the best fit values of 8 SNe1a parameters, along with Ω_m, Ω_Λ, w, w_0 and w_1. This method has the advantage of being bias-free as compared to the conventional least χ^2 method, as arbitrary values of uncertainty σ can be added to each data point in the χ^2 calculation to make the data fit a desired model.

For the ΛCDM model, we obtain best fit values of $\Omega_m = 0.40$, $\Omega_\Lambda = 0.55$, which is within 1σ of N16. When we do a model comparison based on calculated AIC and BIC values, we see that the ΛCDM model is still favoured over the wCDM model and its parametrizations.

With the MLE and MCMC method, our contour plots for all 5 models cross the line of no acceleration, which is a different result from that obtained by R98 during their discovery of the Universe's accelerating expansion, and is similar to the contour plot published by N16. Moreover, when we generalize the ΛCDM model to the wCDM model and its parametrizations, we see that constant expansion becomes increasingly favoured (the contours cross the line of no acceleration at smaller σ values). The conclusion of an accelerating expansion purely based on the study of SNe1a data does not seem to be as secure as before, thus more work needs to be done for us to better understand the Universe we live in.

References

[1] A.G. Riess, A.V. Filippenko, P. Challis, A. Clocchiatti, A. Diercks, P.M. Garnavich, R.L. Gilliland, C.J. Hogan, S. Jha, R.P. Kirshner et al., The Astronomical Journal **116**, 1009 (1998)

[2] S. Perlmutter, G. Aldering, G. Goldhaber, R.A. Knop, P. Nugent, P.G. Castro, S. Deustua, S. Fabbro, A. Goobar, D.E. Groom et al., The Astrophysical Journal **517**, 565 (1999)

[3] J.A. Frieman, B. Bassett, A. Becker, C. Choi, D. Cinabro, F. DeJongh, D.L. Depoy, B. Dilday, M. Doi, P.M. Garnavich et al., Astronomical Journal **135**, 338 (2008), 0708.2749

[4] A. Conley, J. Guy, M. Sullivan, N. Regnault, P. Astier, C. Balland, S. Basa, R.G. Carlberg, D. Fouchez, D. Hardin et al., Astrophys. J. Suppl. **192**, 1 (2011), 1104.1443

[5] A.G. Riess et al., Astrophys. J. **659**, 98 (2007), astro-ph/0611572

[6] G. Miknaitis et al., Astrophys. J. **666**, 674 (2007), astro-ph/0701043

[7] M. Ganeshalingam, W. Li, A.V. Filippenko, Monthly Notices of the Royal Astronomical Society **433**, 2240 (2013), http://oup.prod.sis.lan/mnras/article-pdf/433/3/2240/13761647/stt893.pdf

[8] M. Hicken, P. Challis, S. Jha, R.P. Kirshner, T. Matheson, M. Modjaz, A. Rest, W.M. Wood-Vasey, G. Bakos, E.J. Barton et al., Astrophys. J. **700**, 331 (2009), 0901.4787

[9] C. Contreras, M. Hamuy, M.M. Phillips, G. Folatelli, N.B. Suntzeff, S.E. Persson, M. Stritzinger, L. Boldt, S. González, W. Krzeminski et al., Astronomical Journal **139**, 519 (2010), 0910.3330

[10] D.M. Scolnic et al., Astrophys. J. **859**, 101 (2018), 1710.00845

[11] D. Brout et al. (DES) (2018), 1811.02377

[12] J.T. Nielsen, A. Guffanti, S. Sarkar, Sci. Rep. **6**, 35596 (2016), 1506.01354

[13] H. Shariff, X. Jiao, R. Trotta, D.A. van Dyk, Astrophys. J. **827**, 1 (2016), 1510.05954

[14] M. Betoule et al. (SDSS), Astron. Astrophys. **568**, A22 (2014), 1401.4064

[15] B.S. Haridasu, V.V. Luković, R. D'Agostino, N. Vittorio, Astron. Astrophys. **600**, L1 (2017), 1702.08244

[16] S. Nesseris, L. Perivolaropoulos, Phys. Rev. **D72**, 123519 (2005), astro-ph/0511040

[17] Y.Y. Xu, X. Zhang, The European Physical Journal C **76**, 588 (2016)

[18] J. Guy et al. (SNLS), Astron. Astrophys. **466**, 11 (2007), astro-ph/0701828

[19] M.G. Kendall, A. Stuart, J.K. Ord, eds., *Kendall's Advanced Theory of Statistics* (Oxford University Press, Inc., New York, NY, USA, 1987), ISBN 0-195-20561-8

[20] G.E.P. Box, G.C. Tiao, Ann. Math. Statist. **36**, 1468 (1965)

[21] D. Rubin, B. Hayden, The Astrophysical Journal **833**, L30 (2016)

[22] M. Pitkin, *Yet another matlab mcmc code* (2015), http://www.astro.gla.ac.uk/~matthew/blog/?p=813

[23] J. Goodman, J. Weare, Commun. Appl. Math. Comput. Sci. **5**, 65 (2010)

[24] H. Akaike, IEEE Transactions on Automatic Control **19**, 716 (1974)

[25] G. Schwarz, Ann. Statist. **6**, 461 (1978)

A Phenomenological Model of Effectively Oscillating Massless Neutrinos and Its Implications

Jianlong Lu[1,*], *Aik Hui* Chan[1], and *Choo Hiap* Oh[1,2]

[1]Department of Physics, National University of Singapore, Singapore
[2]Centre for Quantum Technologies, National University of Singapore, Singapore

Abstract. We discuss an alternative picture of neutrino oscillation. In this phenomenological model, the flavor-changing phenomena of massless neutrinos arise from scattering processes between neutrinos and four types of undetected spin-0 massive particles pervading throughout the Universe, instead of neutrinos' own nature. These scattering processes are kinematically similar to Compton scattering. One type of left-handed massless sterile neutrino is needed in order to reproduce the neutrino oscillation modes predicted in the theory of neutrino mixing. Implications of our model include the existence of sterile neutrinos, the nonconservation of active neutrinos, the possible mismatch among three neutrino mass squared differences Δm^2_{ij} interpreted in the theory of neutrino mixing, the spacetime dependence of neutrino oscillation, and the impossibility of neutrinoless double beta decay. Several important open problems in neutrino physics become trivial or less severe in our model, such as the smallness of neutrino masses, neutrino mass hierarchy, the mechanism responsible for neutrino masses, and the Dirac/Majorana nature of neutrinos.

1 Introduction

The concept of neutrinos historically originated from Wolfgang Pauli's proposal in 1930 to explain the continuous energy spectra of electrons produced in beta decay [1]. In the Standard Model (SM) of particle physics [2], neutrinos are described together with the left-handed components of charged leptons as doublets of the $SU(2)_L$ group,

$$l_L = \begin{pmatrix} \nu_{eL} \\ e_L \end{pmatrix}, \begin{pmatrix} \nu_{\mu L} \\ \mu_L \end{pmatrix}, \begin{pmatrix} \nu_{\tau L} \\ \tau_L \end{pmatrix}. \tag{1}$$

Each of the right-handed components of charged leptons (e_R, μ_R, τ_R) forms singlet of $SU(2)_L$ by itself. All interactions involving neutrinos in the SM can be described by the following charged current (CC) interaction and neutral current (NC) interaction Lagrangian densities [3],

$$-\mathcal{L}_{CC} = \frac{g}{\sqrt{2}} \sum_{l=e,\mu,\tau} \bar{\nu}_{lL} \gamma^\mu l_L^- W_\mu^+ + \text{h.c.}\,, \tag{2}$$

$$-\mathcal{L}_{NC} = \frac{g}{2\cos\theta_W} \sum_{l=e,\mu,\tau} \bar{\nu}_{lL} \gamma^\mu \nu_{lL} Z_\mu^0, \tag{3}$$

*e-mail: jianlong_lu@u.nus.edu

in which θ_W is the Weinberg angle and g is the SU(2) coupling coefficient.
It is assumed that no right-handed components of neutrinos exist in the SM, which stops us from writing down the neutrino Dirac mass term $\mathcal{L}^D = -m_D(\bar{\nu}_R\nu_L + \bar{\nu}_L\nu_R)$ where $m_D = \frac{yv}{\sqrt{2}}$ is determined by the neutrino Yukawa coupling coefficient y and vacuum expectation value of Higgs field $\frac{v}{\sqrt{2}}$ [4]. Moreover, a Majorana mass term will not be $SU(2)_L$ invariant and cannot be generated within the SM containing only the Higgs doublet [5]. Hence, neutrinos must be massless in the SM if no further extensions or modifications are introduced.
However, from the late 1960s, a number of observations convincingly suggest that there exist flavor-changing phenomena of neutrinos during their propagation in spacetime, later called neutrino oscillation, which are obviously not predicted by the SM. One example is the well-known Homestake experiment headed by Raymond Davis Jr. and John N. Bahcall [6]. In Homestake experiment, the electron neutrinos coming from ^{8}B decay in the Sun were detected and counted, which revealed a significant deficit compared with the theoretical prediction based on the standard solar model and gave rise to the solar neutrino problem. At the present stage, the most widely accepted solution to this dilemma is to introduce a unitary matrix U called Pontecorvo–Maki–Nakagawa–Sakata matrix (PMNS matrix) which describes the mismatch of neutrino mass eigenstates and neutrino flavor eigenstates [7]. For three-neutrino case, the neutrino mixing is encoded in the following equation,

$$\begin{pmatrix} \nu_e \\ \nu_\mu \\ \nu_\tau \end{pmatrix} = U \begin{pmatrix} \nu_1 \\ \nu_2 \\ \nu_3 \end{pmatrix}, \tag{4}$$

where $(\nu_e, \nu_\mu, \nu_\tau)$ are neutrino flavor eigenstates and (ν_1, ν_2, ν_3) are neutrino mass eigenstates. There are many different ways to parametrize PMNS matrix. One of the commonly used parameterizations of PMNS matrix, the so-called standard parameterization, is as follows [3]:

$$U = \begin{pmatrix} c_{12}c_{13} & s_{12}c_{13} & s_{13}e^{-i\delta} \\ -c_{23}s_{12} - s_{23}c_{12}s_{13}e^{i\delta} & c_{23}c_{12} - s_{23}s_{12}s_{13}e^{i\delta} & s_{23}c_{13} \\ s_{23}s_{12} - c_{23}c_{12}s_{13}e^{i\delta} & -s_{23}c_{12} - c_{23}s_{12}s_{13}e^{i\delta} & c_{23}c_{13} \end{pmatrix} \begin{pmatrix} e^{i\phi_1} & & \\ & e^{i\phi_2} & \\ & & 1 \end{pmatrix}, \tag{5}$$

where $s_{ij} \equiv \sin\theta_{ij}$ and $c_{ij} \equiv \cos\theta_{ij}$. There are six parameters in the standard parameterization, including three neutrino mixing angles $(\theta_{12}, \theta_{13}, \theta_{23})$ and three CP-violating phases (δ, ϕ_1, ϕ_2). Without loss of generality, we may assume that $\theta_{ij} \in [0, \frac{\pi}{2}]$ for all neutrino mixing angle θ_{ij}. The range of three CP-violating phases is $[0, 2\pi]$. The first CP-violating phase δ is called the Dirac phase, upon which the CP-violating effects and T-violating effects of neutrino oscillation depend. The remaining two CP-violating phase ϕ_1 and ϕ_2 are called the Majorana phases that are irrelevant in neutrino oscillation phenomena and needed only when massive neutrinos are Majorana particles [3][8]. It is assumed in this paper that there are three and only three types of light left-handed active neutrinos, since this is suggested by the measurement of the total decay width of the Z^0 bosons [3].
The following neutrino flavor transition probability formulas in the theory of neutrino mixing are easily derived [2],

$$P^{\text{massive}}_{\nu_\alpha \to \nu_\beta,\ \alpha \neq \beta} = \sum_{i=1}^{3} |U_{\alpha i}|^2 |U_{\beta i}|^2 + 2\mathrm{Re}\Big[\sum_{j<k} U_{\alpha j} U_{\beta k} U^*_{\alpha k} U^*_{\beta j} e^{-i\frac{\Delta m^2_{kj} L}{2E}}\Big] \tag{6}$$

where three neutrino mass squared differences $\Delta m^2_{kj} = m^2_k - m^2_j$ are defined. $U_{\alpha i}$ represent entries of PMNS matrix with $\alpha = e, \mu, \tau$ and $i = 1, 2, 3$. Note that the rows of PMNS matrix are labeled by three neutrino flavors (e, μ, τ).

Although this three-neutrino mixing scheme neatly explains the observed flavor-changing phenomena of neutrinos, it is still accompanied by some unsolved problems, both on the theoretical side and on the experimental side. We need a mechanism that generates the tiny neutrino masses, which requires appropriate extension in the SM such as existence of heavy neutrinos in the seesaw mechanism [9]. The absolute values of neutrino masses are also yet to be confirmed, for which we can only give nonzero upper bounds inferred from cosmological observations (such as cosmic microwave background and Type Ia Supernovae) [10], energy of electrons from nuclear beta decay [11], and the half life of neutrinoless double beta decay ($0\nu\beta\beta$) [12]. We do not know whether massive neutrinos are Dirac particles or Majorana particles, awaiting hints from neutrinoless double beta decay that can happen only if massive neutrinos are Majorana particles [13]. The list goes on. There are much more problems existing in our way to a full understanding of massive neutrinos.
Although the neutrino mixing model is sufficient to produce flavor-changing phenomena of neutrinos, it is logically not a necessary condition. The possibility of massless neutrinos is still not ruled out because the direct hint about the absolute scale of neutrino masses currently known is upper bound instead of nonzero lower bound. Note that three nonzero neutrino mass squared differences Δm^2_{ij}, appearing in the neutrino transition probability formulas mentioned above, are model-dependent thus cannot be used to rule out massless neutrinos. From the next section, we look at the flavor-changing behavior of neutrinos from another point of view. This trial is not meant to challenge the mainstream theory of neutrino mixing, but to explore some new possibilities.

2 The Model

Neutrinos (anti-neutrinos) are understood as massless left-handed (right-handed) particles in our phenomenological model, whose flavor-changing behaviors are due to their interactions with some undetected particles pervading throughout the Universe instead of their own intrinsic properties. In the following analysis, the formulas of neutrino transition probabilities in our model are given, which mathematically have the same form as those in the theory of neutrino mixing.
We propose that there are 4 types of undetected massive spin-0 particles filling the Universe together with their anti-particles. Each particle carries a specific combination of three leptonic numbers. For convenience of reference, we call these particles R and their anti-particles $\bar{R}$. The letter R stands for *reservoir*, since these undetected particles together act as a reservoir of leptonic numbers. Subscripts are introduced to specify the leptonic numbers of each R. An R_{ab} carries one unit of a-leptonic number and one unit of b-leptonic number, which are opposite to those of $\bar{R}_{ab}$. For example, the electron number, muon number, and tau number of $R_{e\mu}$ are respectively +1, +1, and 0. Besides $R_{e\mu}$, $R_{e\tau}$, and $R_{\mu\tau}$, which carry two nonzero leptonic numbers, there also exists $R_{e\mu\tau}$. In our model, it is these particles that exchange leptonic numbers with massless neutrinos and induce the neutrino flavor-changing phenomena, somewhat similar to the change of color charges of quarks caused by gluons.
For convenience of reference, we list and label all 12 flavor-changing processes in **Table 1**. The labels are used as subscripts for the corresponding cross sections. For simplicity instead of theoretical necessity, we assume that each type of R has the same number density distribution in spacetime, denoted by η. Note that η is a function of $x \equiv (t, \mathbf{x})$, which is generally not constant since R at least takes part in gravitational interaction. It is easy to notice that all 12 processes listed in **Table 1** respect the conservation of total number of neutrinos including sterile neutrino ν_s, which is important in our model to reproduce the neutrino transition probability formulas in the theory of neutrino mixing.

To ensure that in each flavor-changing process the outgoing neutrino and the incoming neutrino move in the same direction, it suffices to assign zero spin to all R. According to the conservation of angular momentum, it is obvious that the intrinsically left-handed neutrinos must keep the direction of propagation unchanged after scattering. Furthermore, we can show that the outgoing neutrino and the incoming neutrino have the same amount of energy. Observe that all 12 processes are similar to Compton scattering, in which a massless photon is scattered by a massive electron. The wavelength shift formula for the scattered photon in Compton scattering is [14]

$$\lambda' - \lambda = \frac{h}{mc}(1 - \cos\theta), \tag{7}$$

where m is the rest mass of the massive target particle; θ is the scattering angle of the massless particle; λ' and λ respectively represent the initial and final wavelengths of the massless particle. In our case, the feature that the outgoing neutrino and the incoming neutrino move in the same direction, i.e., $\theta = 0$, implies

$$\lambda' - \lambda = 0. \tag{8}$$

It means that the outgoing neutrino has the same energy as the incoming neutrino. Note that Compton's formula works only in the inertial frame of reference in which the target particle is initially at rest. However, since the outgoing neutrino and the incoming neutrino move in the same direction, the energy equality will be invariant under any Lorentz transformation.
In the following analysis, ρ_{ν_α} with $\alpha = e, \mu, \tau, s$ represent the normalized neutrino number densities such that $\sum_\alpha \rho_{\nu_\alpha} = 1$; σ_j with $j = 1, 2, ..., 12$ are the cross sections of those 12 processes listed in **Table 1**. The flavor-changing behavior of neutrinos is described by the following differential equation:

$$\frac{d}{dt}\begin{pmatrix}\rho_{\nu_e}\\ \rho_{\nu_\mu}\\ \rho_{\nu_\tau}\\ \rho_{\nu_s}\end{pmatrix} = A\eta\begin{pmatrix}-(\sigma_1+\sigma_2+\sigma_7) & \sigma_3 & \sigma_5 & \sigma_8\\ \sigma_1 & -(\sigma_3+\sigma_4+\sigma_9) & \sigma_6 & \sigma_{10}\\ \sigma_2 & \sigma_4 & -(\sigma_5+\sigma_6+\sigma_{11}) & \sigma_{12}\\ \sigma_7 & \sigma_9 & \sigma_{11} & -(\sigma_8+\sigma_{10}+\sigma_{12})\end{pmatrix}\begin{pmatrix}\rho_{\nu_e}\\ \rho_{\nu_\mu}\\ \rho_{\nu_\tau}\\ \rho_{\nu_s}\end{pmatrix}, \tag{9}$$

where A is some real constant. For convenience of reference, we name the 4×4 matrix in the above equation by M.
It is easy to see that one eigenvalue of M is 0, which arises from the conservation of neutrinos encoded in M. It is this zero eigenvalue that contributes the constant term in each neutrino transition probability formula. The other three eigenvalues of M are generally nonzero. To reproduce the nontrivial oscillating behaviour predicted in the theory of neutrino mixing, all the remaining three eigenvalues must be nonzero and purely imaginary. It can be easily achieved by assuming that M is skew-Hermitian, i.e., $M^\dagger = -M$. This condition greatly reduces the degrees of freedom in our model and simplifies M as follows,

$$M = \begin{pmatrix}-(\sigma_1+\sigma_2+\sigma_7) & -\sigma_1^* & -\sigma_2^* & -\sigma_7^*\\ \sigma_1 & \sigma_1^* - \sigma_4 - \sigma_9 & -\sigma_4^* & -\sigma_9^*\\ \sigma_2 & \sigma_4 & \sigma_2^* + \sigma_4^* - \sigma_{11} & -\sigma_{11}^*\\ \sigma_7 & \sigma_9 & \sigma_{11} & \sigma_7^* + \sigma_9^* + \sigma_{11}^*\end{pmatrix}. \tag{10}$$

We denote four eigenvalues of M by $(\lambda_1, \lambda_2, \lambda_3, \lambda_4)$, including $\lambda_1 = 0$. All four eigenvalues can be directly calculated and explicitly given. Note that in our equation cross sections are living in the complex plane instead of the real axis. For a square matrix with all entries being real, any purely imaginary eigenvalue must have a complex conjugate partner. Since one eigenvalue of our 4×4 matrix M is zero, it is impossible for M to have three nonzero purely imaginary eigenvalues if all cross sections are real numbers. There are some important

features about the relations among eigenvalues of M:
1. All nonzero eigenvalues are purely imaginary as M is skew-Hermitian. They correspond to three frequencies of neutrino transition probabilities.
2. In general, eigenvalues of M are not degenerate.
3. There exist parameters $(\sigma_1, \sigma_2, \sigma_4, \sigma_7, \sigma_9, \sigma_{11})$ such that $(\lambda_2 + \lambda_3 + \lambda_4)$, $(\lambda_2 + \lambda_3 - \lambda_4)$, $(\lambda_2 - \lambda_3 + \lambda_4)$, and $(-\lambda_2 + \lambda_3 + \lambda_4)$ are simultaneously nonzero.
For each eigenvalue, we can explicitly determine its eigenvector up to a complex scale factor. All degrees of freedom hidden in the four complex scale factors can be completely eliminated by imposing initial condition. We represent the four eigenvectors as the following complex column matrices:

$$\begin{pmatrix}\alpha_1\\ \alpha_2\\ \alpha_3\\ \alpha_4\end{pmatrix}, \begin{pmatrix}\beta_1\\ \beta_2\\ \beta_3\\ \beta_4\end{pmatrix}, \begin{pmatrix}\gamma_1\\ \gamma_2\\ \gamma_3\\ \gamma_4\end{pmatrix}, \begin{pmatrix}\delta_1\\ \delta_2\\ \delta_3\\ \delta_4\end{pmatrix}, \tag{11}$$

which satisfy the initial condition physically corresponding to a beam of neutrinos that are all electron neutrinos at $t = 0$,

$$\begin{pmatrix}\alpha_1\\ \alpha_2\\ \alpha_3\\ \alpha_4\end{pmatrix} + \begin{pmatrix}\beta_1\\ \beta_2\\ \beta_3\\ \beta_4\end{pmatrix} + \begin{pmatrix}\gamma_1\\ \gamma_2\\ \gamma_3\\ \gamma_4\end{pmatrix} + \begin{pmatrix}\delta_1\\ \delta_2\\ \delta_3\\ \delta_4\end{pmatrix} = \begin{pmatrix}1\\ 0\\ 0\\ 0\end{pmatrix}. \tag{12}$$

Then the solution to the flavor-changing equation is

$$\begin{pmatrix}\rho_{\nu_e}\\ \rho_{\nu_\mu}\\ \rho_{\nu_\tau}\\ \rho_{\nu_s}\end{pmatrix} = \begin{pmatrix}\alpha_1\\ \alpha_2\\ \alpha_3\\ \alpha_4\end{pmatrix} e^{A\eta\lambda_1 t} + \begin{pmatrix}\beta_1\\ \beta_2\\ \beta_3\\ \beta_4\end{pmatrix} e^{A\eta\lambda_2 t} + \begin{pmatrix}\gamma_1\\ \gamma_2\\ \gamma_3\\ \gamma_4\end{pmatrix} e^{A\eta\lambda_3 t} + \begin{pmatrix}\delta_1\\ \delta_2\\ \delta_3\\ \delta_4\end{pmatrix} e^{A\eta\lambda_4 t}. \tag{13}$$

Note that, although the entries of column vector $\begin{pmatrix}\rho_{\nu_e} & \rho_{\nu_\mu} & \rho_{\nu_\tau} & \rho_{\nu_s}\end{pmatrix}^T$ are all real number at $t = 0$, complex entries will arise from time evolution. Fortunately, this problem can be easily solved if we focus on the real parts. Observe that the matrix M contains the conservation rule

$$\frac{d}{dt} \sum_{\alpha=e,\mu,\tau,s} \rho_{\nu_\alpha} = 0, \tag{14}$$

which immediately implies

$$\frac{d}{dt}\mathrm{Re}\Big(\sum_{\alpha=e,\mu,\tau,s} \rho_{\nu_\alpha}\Big) = \frac{d}{dt} \sum_{\alpha=e,\mu,\tau,s} \mathrm{Re}(\rho_{\nu_\alpha}) = 0. \tag{15}$$

We interprete $\mathrm{Re}(\rho_{\nu_\alpha})$ as the physical normalized number density of ν_α . Now we can write down the neutrino transition probability formulas in our model. Note that we have chosen the initial condition such that the neutrinos are all initially electron neutrinos. The transition probability formulas for processes starting from muon neutrinos and tau neutrinos can be easily obtained by imposing appropriate initial conditions.
The transition probabilities for $\nu_e \to \nu_\mu$, $\nu_e \to \nu_\tau$, and $\nu_e \to \nu_s$ are respectively

$$P^{\mathrm{massless}}_{\nu_e\to\nu_\mu} = \mathrm{Re}(\alpha_2) + \mathrm{Re}(\beta_2 e^{A\eta\lambda_2 t}) + \mathrm{Re}(\gamma_2 e^{A\eta\lambda_3 t}) + \mathrm{Re}(\delta_2 e^{A\eta\lambda_4 t}), \tag{16}$$

$$P^{\mathrm{massless}}_{\nu_e\to\nu_\tau} = \mathrm{Re}(\alpha_3) + \mathrm{Re}(\beta_3 e^{A\eta\lambda_2 t}) + \mathrm{Re}(\gamma_3 e^{A\eta\lambda_3 t}) + \mathrm{Re}(\delta_3 e^{A\eta\lambda_4 t}), \tag{17}$$

$$P^{\text{massless}}_{\nu_e \to \nu_s} = \text{Re}(\alpha_4) + \text{Re}(\beta_4 e^{A\eta\lambda_2 t}) + \text{Re}(\gamma_4 e^{A\eta\lambda_3 t}) + \text{Re}(\delta_4 e^{A\eta\lambda_4 t}). \tag{18}$$

One remarkable property of the above formulas is that they mathematically have the same form as their counterparts in the theory of neutrino mixing. As we can notice, the phase factors $\text{Im}(A\eta\lambda_i t)$ with $i = 2, 3, 4$ in our model do not necessarily match the physical interpretation of the phase factors $-\frac{\Delta m^2_{kj} L}{2E}$ with $1 \leq j < k \leq 3$ in the theory of neutrino mixing, due to the third feature of eigenvalues of M mentioned previously. Another crucial difference between our model and the theory of neutrino mixing is that in our model the total number of active neutrinos (ν_e, ν_μ, ν_τ) is not conserved during time evolution. The sterile neutrino ν_s is not introduced for fun, but necessary for the existence of three nontrivial oscillation frequencies in the neutrino transition probabilities.

We would like to remark that, as a special case of our model, the relation among three oscillation frequencies in the theory of neutrino mixing can emerge in our model if we simply demand that all diagonal entries of skew-Hermitian M vanish. This condition further simplifies matrix M as follows:

$$M = \begin{pmatrix} 0 & -\sigma_1^* & -\sigma_2^* & \sigma_1^* + \sigma_2^* \\ \sigma_1 & 0 & -\sigma_4^* & -\sigma_1 + \sigma_4^* \\ \sigma_2 & \sigma_4 & 0 & -\sigma_2 - \sigma_4 \\ -\sigma_1 - \sigma_2 & \sigma_1^* - \sigma_4 & \sigma_2^* + \sigma_4^* & 0 \end{pmatrix} \tag{19}$$

with $\sigma_1 + \sigma_2 + \sigma_4 \in \mathbb{R}$. The nonzero eigenvalues of this matrix satisfy

$$\lambda_2 + \lambda_3 + \lambda_4 = 0. \tag{20}$$

There is no need to worry about any minus sign appearing in the oscillation phase factors since they can be absorbed by the coefficients in front of corresponding sine functions.

We would like to make a remark for the above analysis. The first point is about some processes which are seemingly possible to take part in neutrino flavor-changing phenomena but unmentioned. By massaging 12 processes in **Table 1**, it seems that there are some other processes, including the following typical examples:

$$\nu_e \to \nu_\mu + R_{e\tau} + \bar{R}_{\mu\tau}\ , \quad \nu_\mu + R_{e\tau} + \bar{R}_{\mu\tau} \to \nu_e\ , \tag{21}$$

$$R_{\mu\tau} \to \nu_\mu + R_{e\tau} + \bar{\nu}_e\ , \quad \nu_\mu + R_{e\tau} + \bar{\nu}_e \to R_{\mu\tau}\ , \tag{22}$$

$$\nu_e + \bar{\nu}_\mu \to \bar{R}_{\mu\tau} + R_{e\tau}\ , \quad \bar{R}_{\mu\tau} + R_{e\tau} \to \nu_e + \bar{\nu}_\mu\ , \tag{23}$$

$$\nu_e + \bar{R}_{e\tau} \to \nu_\mu + \bar{R}_{\mu\tau}\ . \tag{24}$$

This list is not arbitrarily given, but has exhausted all possible types of processes we can obtain from playing with the processes in **Table 1**. Since our analysis in the previous section is based only on the processes in **Table 1**, it is necessary to give reasons for ignoring other possibilities, which are as follows:

(1) $\nu_e \to \nu_\mu + R_{e\tau} + \bar{R}_{\mu\tau}$ represents those processes in which one neutrino (anti-neutrino) becomes one neutrino (anti-neutrino), one R, and one $\bar{R}$. These processes are kinematically forbidden in our model since all R and $\bar{R}$ have nonzero rest mass m. Similarly, we can see that the inverse processes of them are also kinematically forbidden.

(2) $R_{\mu\tau} \to \nu_\mu + R_{e\tau} + \bar{\nu}_e$ represents those processes in which one R ($\bar{R}$) becomes one R ($\bar{R}$), one neutrino, and one anti-neutrino. In the center-of-momentum frame, it is obvious that such processes and their inverse processes are all kinematically forbidden.

(3) $\nu_e + \bar{\nu}_\mu \to \bar{R}_{\mu\tau} + R_{e\tau}$ represents those processes in which a pair of neutrino and anti-neutrino with different flavors annihilate and produce a pair of R and $\bar{R}$ with different combinations of

https://doi.org/10.1051/epjconf/202024002002

Table 1. Labels of neutrino flavor-changing processes

Label	Process
1	$\nu_e + R_{\mu\tau} \rightarrow \nu_\mu + R_{e\tau}$
2	$\nu_e + R_{\mu\tau} \rightarrow \nu_\tau + R_{e\mu}$
3	$\nu_\mu + R_{e\tau} \rightarrow \nu_e + R_{\mu\tau}$
4	$\nu_\mu + R_{e\tau} \rightarrow \nu_\tau + R_{e\mu}$
5	$\nu_\tau + R_{e\mu} \rightarrow \nu_e + R_{\mu\tau}$
6	$\nu_\tau + R_{e\mu} \rightarrow \nu_\mu + R_{e\tau}$
7	$\nu_e + R_{\mu\tau} \rightarrow \nu_s + R_{e\mu\tau}$
8	$\nu_s + R_{e\mu\tau} \rightarrow \nu_e + R_{\mu\tau}$
9	$\nu_\mu + R_{e\tau} \rightarrow \nu_s + R_{e\mu\tau}$
10	$\nu_s + R_{e\mu\tau} \rightarrow \nu_\mu + R_{e\tau}$
11	$\nu_\tau + R_{e\mu} \rightarrow \nu_s + R_{e\mu\tau}$
12	$\nu_s + R_{e\mu\tau} \rightarrow \nu_\tau + R_{e\mu}$

leptonic numbers. According to the conservation of angular momentum, the incoming pair of neutrino and anti-neutrino must move in the same direction so that the total spin angular momentum vanishes. Then there exists an inertial frame of reference in which the total energy of them is less than $2mc^2$. Hence, this type of processes is kinematically forbidden. It is easily understood that the same conclusion can be made for their inverse processes.
(4) $\nu_e + \bar{R}_{e\tau} \rightarrow \nu_\mu + \bar{R}_{\mu\tau}$ represents those processes in which one neutrino (anti-neutrino) and one $\bar{R}$ (R) interact and exchange leptonic numbers. For such processes only the incoming neutrino and the outgoing neutrino are indirectly known, therefore these processes are indistinguishable from those processes listed in **Table 1**. We can effectively regard them as forbidden processes and own their contributions all to corresponding processes listed in **Table 1**.
Another assumption we have implicitly made in the analysis of our model is that the number density η is treated as a constant independent of position in spacetime while solving differential equation, which is generally not true since the massive particles R (and their anti-particles $\bar{R}$) at least take part in the gravitational interaction. This assumption should be relaxed when the propagation distance of neutrinos is comparable with the size of large scale structure of the Universe and when the propagation time of neutrinos is comparable with the age of the Universe. As particles that only participate in the neutrino flavor-changing scattering processes and gravitational interaction, R and $\bar{R}$ can be deemed to be new candidates of dark matters.

3 Discussion

Besides reproducing the neutrino transition probability formulas, our phenomenological model gives the following implications showing essential difference from the theory of neutrino mixing.
The first implication is the existence of exactly one type of sterile neutrino coupling with active neutrinos via Compton-like scattering processes involving R and $\bar{R}$. It is the experimental verification of the existence of three nonzero oscillation frequencies in the neutrino transition probabilities, interpreted as $\frac{\Delta m_{kj}^2}{2E}$ in the theory of neutrino mixing, that requires the dimension of matrix M in our model to be 4×4. In contrast, if neutrinos are allowed to oscillate among sterile type and active types in the theory of neutrino mixing, there will be more than three oscillation frequencies in the neutrino transition probabilities. Since in our model active neutrinos can become sterile neutrinos during propagation, the second implication is obvious: the total number of active neutrinos during propagation is generally not conserved. This implication can in principle be tested by precise measurement of total flux of active neutrinos from the Sun or nuclear reactors. The third implication is about the relation among three oscillation frequencies, which in the theory of neutrino mixing is inferred from the interpretation of oscillation frequencies $\frac{\Delta m_{kj}^2}{2E}$ and the self-evident equality $\Delta m_{31}^2 = \Delta m_{32}^2 + \Delta m_{21}^2$. As mentioned previously, for the skew-Hermitian matrix M in our model, there exist parameters $(\sigma_1, \sigma_2, \sigma_4, \sigma_7, \sigma_9, \sigma_{11})$ such that $(\lambda_2+\lambda_3+\lambda_4)$, $(\lambda_2+\lambda_3-\lambda_4)$, $(\lambda_2-\lambda_3+\lambda_4)$, and $(-\lambda_2+\lambda_3+\lambda_4)$ are simultaneously nonzero. The fourth implication is that neutrino oscillation has nontrivial spacetime dependence, determined by the distribution of R and $\bar{R}$ in the Universe, which is nonnegligible for long-range propagation of neutrinos. This property may shed some light on the role of dark matters in the large scale structure of the Universe. Furthermore, the unusual complex nature of scattering cross sections in our model may hint at something deeper behind the flavor-changing scattering processes.
It is worth noticing that the smallness of neutrino masses and the neutrino mass hierarchy are no longer problems in our model since neutrinos return to their roles in the SM as massless fermions. Moreover, these massless neutrinos are Dirac particles instead of Majorana particles because they have chirality opposite to their corresponding anti-particles, which implies the impossibility of neutrinoless double beta decay.

4 Conclusion

We have given a phenomenological model in which neutrinos (anti-neutrinos) are all massless left-handed (right-handed) particles as assumed in the SM. There are totally three types of active neutrinos (ν_e, ν_μ, ν_τ) and one type of sterile neutrino (ν_s) in our model, which interact with four types of undetected spin-0 massive particles via Compton-like scattering processes and change flavors during propagation. Although the neutrino transition probability formulas in our model are similar to those in the theory of neutrino mixing, two models are essentially different from each other, as shown by the implications of our model mentioned above.

Acknowledgement

This work is supported by National University of Singapore Research Scholarship.

References

[1] W. Pauli, Phys. Today **31N9**, 27 (1978)

[2] Z. Z. Xing, S. Zhou, *Neutrinos in Particle Physics, Astronomy and Cosmology* (Zhejiang University Press, Hangzhou and Springer-Verlag Berlin Heidelberg, 2011)
[3] M. Tanabashi *et al.* (Particle Data Group), Phys. Rev. D **98**, 030001 (2018)
[4] C. Giunti, M. Laveder, https://arxiv.org/abs/hep-ph/0310238v2 (2004)
[5] Z. Z. Xing, Int. J. Mod. Phys. A **23**, 4255-4272 (2008)
[6] J. N. Bahcall, R. David. Jr., Science **191**, 264-267 (1976)
[7] Z. Maki, M. Nakagawa, S. Sakata, Progress of Theoretical Physics **28**, 870-880 (1962)
[8] C. Giunti, Phys. Lett. B **686**, 41-43 (2010)
[9] O. G. Miranda, J. W. F. Valle, Nucl. Phys. B **908**, 436-455 (2016)
[10] S. R. Choudhury, S. Choubey, JCAP 1809 (2018) 017
[11] M. Aker, *et al.* (KATRIN Collaboration), Phys. Rev. Lett. **123**, 221802 (2019)
[12] J.J. Gómez-Cadenas, J. Martín-Albo, https://arxiv.org/abs/1502.00581v2 (2015)
[13] A. Giuliani, A. Poves, Adv. High Energy Phys. **2012** (2012)
[14] A. H. Compton, Phys. Rev. **21** (5), 483–502 (1923)

Analytical treatment of small scales matter power spectrum in coupled scalar field (CSF) cosmology

Fargiza A. M. Mulki[1,*], and *Hesti* Wulandari[1]

[1]Astronomy Research Division, Faculty of Mathematics and Natural Sciences, Institut Teknologi Bandung, Jl. Ganesha 10, Bandung 40132, Indonesia

Abstract. In this paper, we reconstruct matter power spectrum of a cosmological model in which coupled scalar field acts as dark energy. The coupled scalar field (CSF) dark energy we propose here is a generalized model of quintessence, k-essence or phantom coupled to dark matter (non-baryonic matter) via a coupling constant. The existence of coupling between dark energy and matter allows energy to transfer between them, which may give rise to different observational signatures, especially at perturbation level, including matter power spectrum. In this work, through analytical exploration we studied that possible signature in matter power spectrum that maybe induced by this mode.

1 Introduction

The standard model of cosmology (the ΛCDM model), is constructed by two dark components of the universe, namely the cosmological constant and cold dark matter. Recent results of Planck CMB mission are in very good agreements with base ΛCDM model [1]. Despite being successful in explaining a large set of cosmological observations, the ΛCDM model suffers from several problems. One of the problems is concerning the interpretation of the cosmological constant. Designating the cosmological constant as vacuum energy stored in spacetime leading to severe fine tuning problem [2]. This serious problem has evoked persistent interests in extending cosmology beyond the standard model, including proposals of dynamically changing scalar field as alternative models of dark energy, such as quintessence, k-essence, and phantom [3 – 10]. Considering that Planck CMB results allow dark energy to enter phantom domain ($\omega_{DE} < -1$), and the fact that the energy density value of dark energy today is similar in order of magnitude with that of matter [11], generalizations of ordinary scalar field into phantom dark energy and uncoupled scalar field into coupled scalar field have been suggested [6,12 – 18].

One way to yield phantom scalar field is by allowing scalar field to have negative kinetic energy, so that its state parameter ω_{ϕ} would exceed the lowest value of ordinary scalar field (see ref. [6]). Different types of couplings have been put forward, one type of coupling that has been put forward is coupling between matter and dark energy via a coupling constant Q

* Corresponding author: fargiza@as.itb.ac.id

[12,13,15,16]. Several authors have explored the effect of such a coupling to the perturbation dynamics of universe [3,4,13,15,16]. In this work, we reconstructed matter power spectrum following CSF-DE and investigated how coupling values can affect the shape of matter power spectrum. Whereas, the inquiry of differences between phantom and ordinary field of CSF-DE has been performed by analyzing CMB model generated by using CAMB codes for various ω_{DE} [19].]. In this work, we have reconstructed matter power spectrum following CSF-DE and investigated how coupling values can affect the shape of the matter power spectrum.

2 The Dynamics of Coupled Scalar Field (CSF-DE)

The model used in this work is in the form of lagrangian:

$$L(X,\phi)=X-V(\phi)-Q\rho_m\phi \tag{1}$$

where Q, ρ_m and ϕ stand for coupling constant, general matter density and scalar field respectively, and $X=-\frac{1}{2}g^{\mu\nu}\nabla_\mu\phi\,\nabla_\mu\phi$ is the kinetic component of scalar field. For lagrangian in eq. (1) we can derive the equation of motion for scalar field in Friedmann universe,

$$\phi''+2\mathrm{H}\phi'+a^2V_{,\varphi}=-Q\rho_m, \tag{2}$$

which is usually known as Klein-Gordon equation. Scalar field coupled to general matter satisfies the condition

$$\nabla_\mu T^\mu_\nu=\pm QT_M\nabla_\nu\phi, \tag{3}$$

where the plus $(+)$ sign refers to matter and radiation, minus $(-)$ sign refers to scalar field, and T_M is the trace of fluid energy-momentum tensor. Complete derivation of the dynamics of the model has been performed by several authors [6,15,17,18], so that here we are only interested in using some results obtained mainly by [15] that are relevant to our objectives. The dynamical equations for scalar field with potential $V=V_0e^{-\lambda\phi}$ are given by [15]

$$\frac{dx}{dN}=-3x+\frac{\sqrt{6}}{2\varepsilon}\lambda V_0y^2+\frac{3}{2}x\left[(1-\omega_m)\varepsilon x^2+(1+\omega_m)(1-V_0y^2)\right]-\frac{\sqrt{6}}{2\varepsilon}Q(1-\varepsilon x^2-V_0y^2),$$

$$\frac{dy}{dN}=-\frac{\sqrt{6}}{2}\lambda xy+\frac{3}{2}y\left[(1-\omega_m)\varepsilon x^2+(1+\omega_m)(1-V_0y^2)\right],$$

where $\varepsilon=+1$ for ordinary field and $\varepsilon=-1$ for phantom fields. Above equations are autonomous equations which have infinite number of solutions that move together passing critical points x and y which correspond to particular epochs of the universe. Here we mention several domination stages which are relevant to our next discussion, and for practical reason we ignore the radiation dominated era. Matter dominated era corresponds to $x=-\sqrt{6}Q/3\varepsilon$, $\Omega_\phi=2Q^2/3\varepsilon$, and $\omega_{eff}=2Q^2/3\varepsilon$, matter scaling era corresponds to $Qx=-\sqrt{6}\omega_{eff}/2$, and scalar field dominated era corresponds to $x=\lambda/\sqrt{6}\varepsilon$, $y=[1/V_0(1-\lambda^2/6\varepsilon)]^{1/2}$, $\Omega_\phi=1$, $\omega_\phi=\omega_{eff}=-1+\lambda^2/3\varepsilon$. x and y are dimensionless variables that are introduced to dynamical analysis in phase space, Ω_ϕ is density parameter of CSF-DE, ω_{eff} is efective equation of state parameter, ω_ϕ is equation of state parameter of CSF-DE, λ and V_0 respectively are constant and initial potential (see for details ref. [15]).

To achieve our goal in constructing cosmological power spectra, we need to carry out analysis of perturbation dynamics. In cosmological perturbation theory, structures can be regarded as small fluctuations in the past that evolves with time. For the reason that the dominant force driving the universe on large scales is gravity, we give perturbation to gravity in which mathematically can be expressed by expanding Einstein field equation into Taylor series written as follows

$$\delta\left[R_{\mu\nu}-\frac{1}{2}g_{\mu\nu}R\right]=\delta\left[8\pi GT_{\mu\nu}\right]. \tag{4}$$

By evaluating and simplifying eq. (4) together with perturbing eq. (3), we can obtain following equation [15]

$$\frac{d^2\delta_m}{dN^2}+\left(\frac{1}{2}-\frac{3}{2}\omega_{eff}+\sqrt{6}Qx\right)\frac{d\delta_m}{dN}-\frac{3}{2}\left(1-\Omega_\phi\right)\left\{1+2Q^2\left(1-\Omega_\phi\right)\right\}\delta_m=0.$$

The solutions are in the form $\delta(a) = C_1 a^{\gamma_1} + C_2 a^{\gamma_2}$, with

$$\gamma_{1,2}=\frac{1}{\varepsilon}\left[\frac{3Q^2}{2}-\frac{\varepsilon}{4}\pm\frac{1}{12}\sqrt{192Q^6-576\varepsilon Q^4+432\varepsilon^2Q^2+3324Q^4-252\varepsilon Q^2+225\varepsilon^2}\right], \tag{5}$$

for matter dominated era, and

$$\gamma_{1,2}=\frac{1}{2}\left[-5+\frac{3\lambda^2}{2\varepsilon}\pm\left(5-\frac{3\lambda^2}{2\varepsilon}\right)\right], \tag{6}$$

for scalar field dominated era.

Based on the results in eq. (5) and (6) we see that, for ordinary fields, the existence of coupling Q triggers additional growth rate so that structures grow faster than in the cases of uncoupled and ΛCDM models. On the other hand, phantom fields with negative kinetic energy invert the flow of energy current, so that structures grow slower than in the cases of uncoupled and ΛCDM models. As for the time in the distant future, both ordinary and phantom fields will break down the existing structures, but with a little distinction between them, i.e. in phantom scenario the decay occurs stronger than as in the ordinary one, indicating an explosive expansion of the background universe or big rip.

3 Matter Power Spectrum

In this section we will discuss the possible signatures that may be generated by CSF-DE model. One of the main promising signatures to investigate is matter power spectrum. Here, we will construct matter power spectrum originally starting from CSF-DE model, in which all the aforementioned equations will analytically be transformed into observable parameters.

3.1 Matter transfer function

To construct matter power spectrum we need a function named transfer function. Transfer function is a function that map perturbation in the past into its value we observe today. The first step we need to do is connecting perturbation variables over spaces x and scales k. Therefore, we need Fourier transform that by definition is expressed as

$$\delta(\vec{x},t)=\frac{V}{(2\pi)^3}\int\delta(\vec{k},t)e^{i(\vec{k}.\vec{x})}d\vec{k},$$

where V is a volume that large enough to respect cosmological principle, and $\delta(\vec{x},t)$ is density contrast which satisfies $\rho(\vec{x},t) = \bar{\rho}(t)(1+\delta(\vec{x},t))$. Statistical properties of perturbation are determined well by infinite number of correlation functions ζ_n that states how the density of a region correlated with those of other region(s), for example two correlation function $\zeta_2 = \langle\delta(x_1)\delta(x_2)\rangle$.

By decomposing correlation function ζ_2 into Fourier space, we obtain

$$\zeta_2(\Delta)=\frac{V}{(2\pi)^3}\int\left|\delta\left(\vec{k}\right)\right|^2 e^{-i\vec{k}.\Delta}d\vec{k}. \tag{7}$$

The average values of amplitude in eq. (7) is known as power spectrum, which mathematically is expressed as:

$$P(k)\equiv\left\langle\left|\delta\left(\vec{k}\right)\right|^2\right\rangle.$$

Next, imagine a universe as an open system which has a set of physical laws and conducts its physical processes that is implicitly contained by a function $T(k,t)$. This function transform input into output. In cosmological context, we may regard all perturbations entered the horizon as inputs, and the perturbation observed today as output. Therefore, we may write

$$\delta(k,t)=T(k,t)\,\delta(k,t_0). \tag{8}$$

Here we consider two cases: perturbations that enter the horizon during radiation dominated era, and those that enter the horizon during matter dominated era. We do not consider all perturbations that entered at dark energy dominated era because we know that from observations there are not structures found as large as the scales of horizon at dark energy era.

For small scales in which perturbation enter the horizon at radiation dominated era and undergo physical processes generated during radiation era, matter era, and finally coupled scalar field era. Therefore, we may write all these processes mathematically as

$$k^{3/2}\delta(k,a=1)\approx\underbrace{\delta(k,a)}_{MD}\times\underbrace{\delta(k,a)}_{RD}k^{3/2}$$

which yields

$$T_S(k)\approx\frac{2a_{eq}^{\gamma}\ln\left(a_{eq}\right)}{a_{eq}^{\gamma}-1}\left(\frac{200\pi\sqrt{\Omega_{R,0}}}{a_{eq}ck}\right)^{\gamma}, \tag{9}$$

$$\gamma=\frac{1}{\varepsilon}\left[\frac{3Q^2}{2}-\frac{\varepsilon}{4}+\frac{1}{12}\sqrt{192Q^6-576\varepsilon Q^4+432\varepsilon^2Q^2+3324Q^4-252\varepsilon Q^2+225\varepsilon^2}\right].$$

where we have used eq. (5) (8), and relation:

$$a=\frac{2\pi H_0\sqrt{\Omega_{R,0}}}{ck}$$

for small scales. For large scales in which perturbation enter the horizon at matter dominated era, it experienced the physical processes generated during matter era and coupled scalar field era. However, since we have no evidences for the existence of structures largest than the horizon size of universe at matter-dark energy equality, we can assume that there is no dynamical evolution of perturbation during scalar field dominated era at linear level. Therefore, we can write the process as follows:

$$k^{3/2}\delta(k,a=1)\approx\underbrace{\delta(k,a)}_{MD}k^{3/2}$$

which yields,

$$T_L(k)\approx 1,$$

means that there is no perturbation change in scalar field domination era in the linear regime.

3.2 Power spectrum on small scales

In many literature, power spectrum is usually expressed as:

$$P(k) = N_s k^{n_s} T^2(k) \tag{10}$$

where N_S is a normalization factor that can be confirmed with CMB power spectrum, and n_S refers to scalar spectral index. S subscription refers to scalar perturbation. Because we focus only on small scales region, N_S maybe different from amplitude of primordial power spectrum A_S. N_S can be obtained by calibrating our calculation to CMB simulation such as CAMB code [19]. From our simulation with CAMB we roughly get $N_S \sim 10^8$. It is also possible that the value of n_S depends on scales, meaning that the observed power spectrum deviates from Harrison-Zeldovich spectrum ($n_S = 1$).To see how n_S could be differ from invarint-scale spectrum value, we plot power spectra with various n_S for $Q = 0$ in Figure 1.

Figure 1 shows that our transfer function model for CSF-DE (ordinary scalar fields, $\varepsilon = +1$) deviates from invariant-scale spectrum expected by standard inflation theory. For power spectrum of eq. (10) and transfer function of eq. (9), it is well confirmed by this result that $n_S = 0$ is more preferable than $n_S = 1$. So we propose a power spectrum in the form of eq. (10) with:

$$n_S \approx \begin{cases} 1 & k \ll k_{eq} \\ 0 & k \gg k_{eq} \end{cases} \tag{11}$$

Eq. (11) appears as a requirement for matching model with data. To acquire appropriate values in the small scale region, we make grids for k values in the range of $0.2 < k < 2.7$, and calibrate our result for $Q = 0$ and $n_S = 0$ with simulation for standard model (ΛCDM) using CAMB code. As the result we obtain $N_S \approx 6.5 \times 10^8$. In Figure 2, we show power spectra on small scales for various Q with $n_S = 0$. We also plot the reference CAMB simulation for appropriate range, and Lyman-α data from [20].

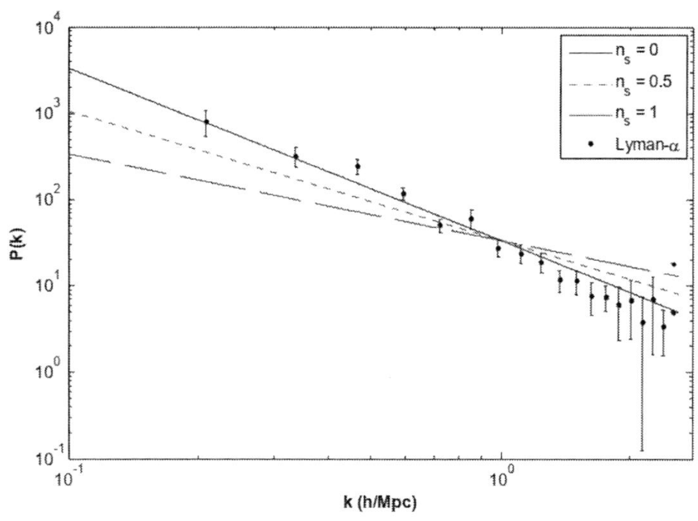

Figure 1. Matter power spectrum on small scales, for various n_S with $Q = 0$ and $N_S = 3 \times 10^8$, compared to Lyman-α data from eBOSS DR14 [20] compiled in repository here[†]

[†] https://github.com/marius311/mpk_compilation/tree/master/dat/lya

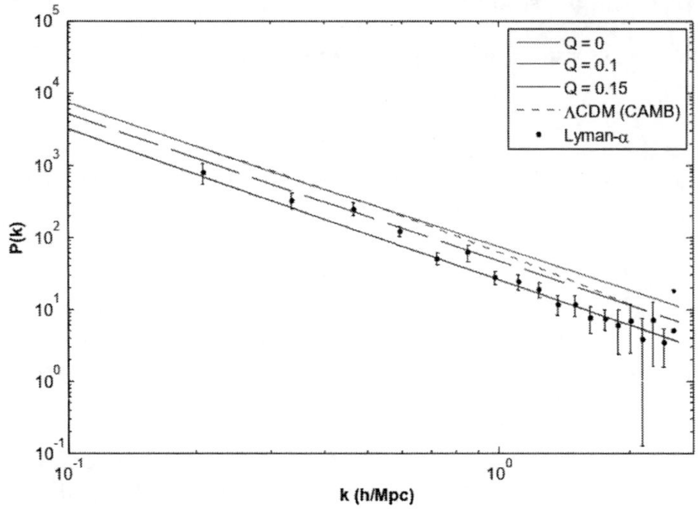

Figure 2. Matter power spectrum on small scales for $n_s = 0$ with various Q compared to ΛCDM model from CAMB and Lyman$-\alpha$ data from [20]

Compared to other models plotted in Figure 1, the power spectrum for $Q = 0.15$, $n_s = 0$ shown in Figure 2 looks fit better with the data. We also find that, $n_s = 0$ with non zero coupling is more preferable by data than the invariant-scale $n_s = 1$. This deviation can be investigated further by numerical study and future data.

4 Conclusion

We have investigated matter power spectrum on small scales stimulated by perturbation during matter dominated era. From analytical investigation, we find that the existence of coupling Q between dark matter and scalar field dark energy affects matter power spectrum profile at small scales and shifts the spectrum to the left with decrement. Considering this result as a promising subject to test CSF-DE model, further study needed to anticipate future data.

References

1. Planck Collaboration, Planck 2018 results. VI. Comological Parameters *arXiv:1807.06209v1* (2018)

2. S. Weinberg, The Cosmological Constant Problem *Rev. Mod. Phys.* **61** 1 (1989)

3. L. Amendola, Phantom Energy Mediates a Long-Range Repulsive Forces *Phys. Rev. Lett.* **93** 181102 (2004)

4. L. Amendola, S. Tsujikawa, and M. Sami, Phantom damping of matter perturbation. *Phys. Lett.* B **632** 155 (2006)

5. R. R. Caldwell, An Introduction to Quintessence *Brazilian Jounal of Physics* **30** 2 (2000)

6. E. J. Copeland, M. Sami, and S. Tsujikawa, Dynamics of Dark Energy *Int. J. Mod. Phys.* D **15** 1753 (2006)

7. F. A. M. Mulki, and H R T Wulandari, Growth of Matter Perturbation in Quintessence Cosmology *American Institute of Physics* **1801** 030002 (2017)

8. B. Ratra, and P. J. E. Peebles, Cosmological consequences of a rolling homogeneous scalar field *Phys. Rev.* D **37** 3406 (1988)

9. M. Sami, and A. Toporensky, Phantom field and the fate of universe. *Mod. Phys. Lett.* **A 19** 1509 (2004)

10. C. Wetterich, The cosmon model for a asymtotically vanishing time dependent cosmological constant *Astron. Astrophys.* **301** 321 (1995)

11. Planck Collaboration, Planck 2015 results. XIV. Dark energy and modified gravity *A & A* A **14** 594 (2016)

12. L. Amendola, Scaling solution in general non–minimal coupling theories *Phys. Rev.* D **60** 043501 (1999)

13. L. Amendola, Coupled Quintessence *Phys. Rev.* D **62** 04351 (2000)

14. B. Gumjudpai, Naskar, and S. Tsujikawa, Coupled Dark Energy: Towards a general description of the dynamics *arXiv:hep-th/0502191v2* (2008)

15. F. A. M. Mulki and H. Wulandari. *J. Phys.: Conf. Ser.* **1354** 012007 (2019)

16. V. Pettorino, Coupled and Extended Quintessence: theoretical differences and structure formation *arXiv: 0802.1086v2* (2008)

17. F. Piazza, and S. Tsujikawa, *JCAP* **0407** 004 (2004)

18. S. Tsujikawa, and M. Sami, A unified approach to scaling solutions in a general cosmological background. *Phys. Lett.* **B 603** 113 (2004)

19. A. Lewis, A. Challinor, and A. Lasenby, *Astrophys. J.* **538** 473 (2000)

20. B. Abolfathi. et al., *ApJS*, **235** 42 (2018)

Astrophysics research at DAP

P.N. Diep[*], *D.T. Hoai, P.T. Nhung, N.T. Phuong, T.T. Thai, P. Tuan-Anh and P. Darriulat*

Department of Astrophysics, Vietnam National Space Center, Vietnam Academy of Science and Technology (VNSC/VAST), 18 Hoang Quoc Viet, Nghia Do, Cau Giay, Hanoi, Vietnam

Abstract. The research and training team of the Department of Astrophysics (DAP), Vietnam National Space Center, is some twenty years old. Currently, we work in the fields of stellar physics and high-redshift galaxies using mm/sub-mm data observed by world-leading radio interferometers such as ALMA and NOEMA. In this report, we present the research team and its activities and give illustrations of the research work with emphasis on the use of archival data.

1 Introduction

DAP, formerly Vietnam Astrophysics Training Laboratory (VATLY), hosts a research team that was put together twenty years ago by Professor Pierre Darriulat at the Institute for Nuclear Science and Technology in Hanoi. Formerly a particle physicist and Research Director of CERN (European Organization for Nuclear Research) he brought to Vietnam detectors and fast electronics units that had been used in particle physics experiments at CERN and had been given to him to help with the development of scientific research in the country. These instruments were old but still working well. In the early years of the laboratory, we used them to set up experiments by ourselves to study cosmic rays which come from space free of charge. By chance, Hanoi sits on the Earth magnetic equator, meaning maximal deflection and therefore higher energy for charged cosmic ray particles. The effect is measured by the so-called rigidity cut-off which is 17 GV in Hanoi, 11 GV in Japan and only 4 GV in Europe and North America. One of our highlight experiments was a series of three measurements of the atmospheric muon flux in Hanoi using a movable home-made telescope. These contributed useful information for neutrino oscillation experiments such as Super-Kamiokande that need to map the muon flux on the Earth surface in order to model what they expect to detect. Indeed both atmospheric muons and neutrinos are decay products of pions in the cascade of secondary particles produced by primary ultra-high energy cosmic rays coming from outer space: muons are therefore tracers of neutrinos. We provided physicists from Super-Kamiokande with our measurements and found them in good agreement with the predictions of their model [1, 2]. Besides carrying experiments by ourselves we also joined the collaboration of the Pierre Auger Observatory (PAO) located in Malargue, in the Argentinean pampa, the largest international collaboration dedicated to the study of extreme energy (extra-galactic) cosmic rays. We took part in many activities of the collaboration: spending time in Malargue to set-up and maintain detectors, to take shifts, to attend collaboration meetings, and most

[*] Corresponding author: pndiep@vnsc.org.vn

importantly to analyse PAO data. Three of us got PhD degrees making our thesis work in the framework of this collaboration. In Hanoi, on the roof of our laboratory, we built a model of the Cherenkov counter of which the PAO array is made, with 1600 such units covering an area of 3000 km^2.

We progressively built up a keen interest for astrophysics. Moreover, as working in a big collaboration was making it difficult for the team to develop an identity of its own, we gradually shifted our interest to radio astronomy where front-line research could be made on studies matching well the size and expertise of a small team. With radio astronomy, the young members of the team were first trained in the laboratory using a pair of Yagi antennas coupled in interferometer mode, a gift of Professor Nguyen Quang Rieu from Paris Observatory. It was used to observe solar interferences and made the subject of a master thesis. Then Pierre bought for us a 2.6 m diameter telescope tuned on the HI line with which we observed the Milky Way, the Sun, the Moon and a few other bright radio sources. It produced nine articles in refereed journals, two master theses and two bachelor dissertations. At the same time we started collaborating with French astrophysicists in Toulouse and Paris on jointly supervised PhD research works. Such collaboration, called "cotutelle" in French, means that for three successive years the PhD student spends part of the year with the Vietnamese team and the other part with the team of the French co-supervisor. The thesis is written in English with substantial summaries in Vietnamese and French, the defence takes place in one of the two countries with a mixed panel and co-chairs from each country and the successful student is awarded both a Vietnamese and a French degree. We were familiar with the scheme for having made use of it three times already with the PAO collaboration. The team includes now three members who obtained their PhD degrees in a cotutelle between a Vietnamese doctoral school and Toulouse University, Paris Observatory and University of Bordeaux. The subjects are in the fields of high red-shift galaxies, evolved stars and star formation respectively. This way of supervising students is truly beneficial both for the PhD students and the whole team, in particular for a young research team like ours. In addition it makes the seed for possible future collaborations.

In January 2015 the whole team moved to the Vietnam National Space Center, which is one of some thirty institutions of the Vietnam Academy of Science and Technology, and became its Department of Astrophysics (DAP).

Currently, astronomy is not included in the curriculum of Vietnamese universities. Only a few universities offer introductory lectures, usually within the frame of the physics departments; an exception is the new University of Science and Technology of Hanoi (USTH) where students can choose to major in astrophysics. We participate in the process of building the teaching programmes in astronomy at USTH, giving lectures, practical classes, and supervising USTH students for their theses and internships at both bachelor and master levels. One of the USTH former master students is staying with us. After having spent six months in France working on MESSEGER data exploring the polar region of planet Mercury for her master thesis, she joined us and we helped her with the completion of the analysis and the redaction of the journal article.

Vietnam recently joined the East Asian Observatory (EAO) as observer. This gave several members of the team the opportunity to join observation programmes of the James Clerk Maxwell Telescope (JCMT) such as a study of magnetic fields in star forming regions (BISTRO) or a survey of nearby evolved stars (NESS). We are taking an active part in the efforts deployed by SEAAN and EAO to foster collaboration in the region, to enhance the quality and quantity of the research work, and to boost training in astronomy and astrophysics.

In Vietnam we devote time and effort to outreach activities aimed at fostering the interest of the public in astronomy and space sciences. In particular we help with the exploitation of VNSC facilities in Hoa Lac, near Hanoi, hosting a 100-seat planetarium, a 50 cm optical telescope and a science museum under current completion. We also take part in training high school students to participate in the International Astronomy and Astrophysics Olympiads (IOAA) and generally in various activities related to public education and training.

Over the past 18 years we have trained 8 PhDs, 10 masters, and 12 bachelor students. We also hosted many interns from both inside Vietnam and abroad. We are the co-authors of 50 SCI articles with the Pierre Auger Collaboration and have published 30 SCI and 31 ASEAN Citation Index articles and conference proceedings of our own. By the end of 2019, the team included ten members (Fig. 1). We publish a Newsletter twice a year where we report on the life of the laboratory and on our research and training activities. More information is available in our web site: https://dap.vnsc.org.vn/.

Fig. 1 The DAP team (October 2019).

2 Research at DAP

Currently, we focus on studies of stellar physics and of far-away galaxies that are observed as they were many Gyr ago. The data which we use are mostly from radio interferometers such as ALMA and NOEMA, the two largest arrays in the southern and northern hemispheres respectively. ALMA stands for Atacama Large Mm/sub-mm Array and NOEMA for NOthern Extended Millimeter Array. ALMA is located in the Atacama desert of the Chilean Andes at 5'000 m above sea level. It consists of 66 antennas of two types, 12 and 7 meters in diameter. NOEMA is in the French Alps on the Plateau de Bure at an altitude of 2'550 m. The complete configuration of NOEMA will include 12 antennas of 15 m diameter by the end of 2021. The signals detected by the antennas of an interferometer array are combined to provide high angular resolution observations of astronomical sources.

Vietnam is not a member of the ALMA and NOEMA partnerships. However, these partnerships, like is often the case in basic research, have a very generous policy aimed in particular at fostering interest for science in developing countries. ALMA make their observations openly accessible to public after a proprietary time of only one year and we are making intensive use of this opportunity. The quality of the data is usually excellent and the quantity available is huge. This open access policy is truly vital for our survival, in particular since our country gives priority to applied rather than basic research and our authorities cannot afford paying important membership fees to international collaborations. As mentioned earlier, we pay much attention to maintaining fruitful contacts with colleagues abroad: such collaboration gives us opportunities not only to enhance our knowledge and skills but also to join them in submitting proposals for observations using facilities to which they have access. Moreover EAO is operating the James Clerk Maxwell Telescope and will take over more telescopes from the region in the future. This gives us access to all the facilities of the Observatory.

A few illustrations of the research work that we are doing are presented below.

2.1 High redshift galaxies

The study of high redshift (high z) galaxies opens a window on the early Universe: they are observed in the state in which they were several Gyr ago. Their study provides important constraints on the mechanism of galaxy formation and insights into physical processes such as star formation in the early Universe. A major difficulty is that such galaxies are very far away and, therefore, very faint. However, this is helped by gravitational lensing: if there is a massive object on the line of sight between the observer and the source, it acts as a lens by bending light rays and an important magnification may result. The mass acting as a lens is usually a galaxy or a cluster of galaxies. The price to pay is that the images are strongly distorted, making the reconstruction of the source brightness distribution difficult. When the lens is essentially a single galaxy one speaks of strong lensing, the configuration that applies to most cases of far-away galaxies that have been studied (see e.g. Blandford et al. [3] or Saha and Williams [4]). We studied in great detail a gravitationally lensed galaxy that hosts a quasar in its centre, RX J0911, at z~2.8, meaning looking back 11.3 Gyr. It is lensed by a foreground galaxy into four images that have been observed by the HST. We studied observations of its gas and dust content, the mm emission of which extends over a large region with very different imaging properties depending on their projected distance to the lens. The problem of de-lensing these observations is difficult and requires, in addition to the knowledge of the quasar HST images, a model of the lens mass distribution [5]. We studied the emission of the CO(7-6) line measured by NOEMA and the continuum emission measured by ALMA. Morpho-kinematics properties of the gas [6] and of the dust [7] have been deduced and possible physical interpretations have been discussed [4]. The gas extends over a distance of 850±120 pc from the centre of the galaxy and its distribution gives evidence for significant ellipticity and velocity gradient caused by molecular outflow and rotation (Fig. 2). The dust component is more compact, by a factor of ~3.4±0.4. These observations make RX J0911 one of the few high-z galaxies for which both dust and gas have been resolved and measured.

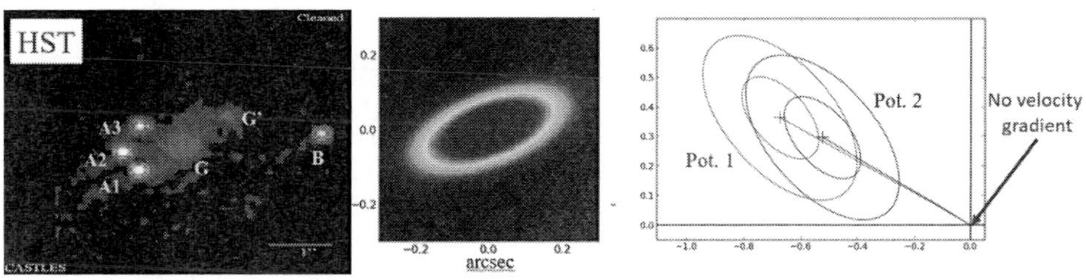

Fig. 2 Left: RX J0911 observed by the Huble Space Telescope. Centre: the gas component revealing ellipticity. Right: Evidence for rotation.

2.2 Evolved stars

The study of evolved star aims at understanding how a star such as the Sun evolves at the end of its lifetime from a nearly perfect spherical shape to the diversity of configurations observed in planetary nebulae. Which are the most important parameters governing the symmetry breaking? Evolved stars often feature a bipolar molecular outflow. A number of evolved stars at different stages of evolution and different orientations in space have been studied at DAP using observations of the high resolution continuum and line emissions from dust grains and gas molecules [8-15]. Methods to study and reconstruct in space axisymmetric radio sources (a common feature of young and old stars) have been developed to this effect [16,17]. A review of studies made by the team in this field, dealing with the presence of high Doppler velocity wings in the spectra of O-rich AGB stars, is presented elsewhere in these Proceedings [18].

Here we show as an example a study of the Red Rectangle using ALMA archival data. It was one of the first stars that we studied after having switched from cosmic rays to radio astronomy [19]. The Red Rectangle is a post-AGB star having its axis perpendicular to the line of sight. It looks like a red rectangle when observed in infrared hence the name. AGB stands for Asymptotic Giant Branch which is a region of the Hertzsprung-Russell diagram populated by cool luminous Red Giants. All low- to intermediate-mass stars (0.6 to 10 solar masses) will go through this phase of stellar evolution. Post-AGB stars are on the way to become planetary nebulae having a degenerate core that will soon be a White Dwarf when what is left of the circumstellar envelope will have completely diluted in the interstellar medium. For the study of the Red Rectangle we used ALMA observations of the CO(6-5) and CO(3-2) emissions of unprecedented quality.

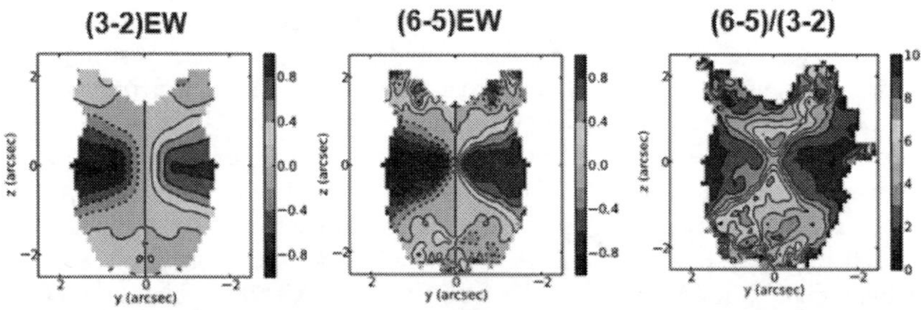

Fig. 3 Red Rectangle: mean Doppler velocity maps (left and centre) and line intensity ratio (right).

Fig. 3 (left and centre) displays the mean Doppler velocity maps for each line separately. The East-West asymmetry reveals clearly the rotation of the equatorial region about the star axis. The map of the ratio of the two line intensities, a proxy of temperature, shows evidence for a biconical structure down to very small distances from the central star (Fig. 3 right). We modelled the gas kinematics dividing the velocity fields in two regions. In the polar regions, the lines of field are meridian parabolas joining smoothly with the equatorial region where trajectories are spirals accounting for both rotation and expansion (Fig. 4 left and central). The best fit (Fig. 4 right) gives a surprisingly good description of the data given the crudeness of the model.

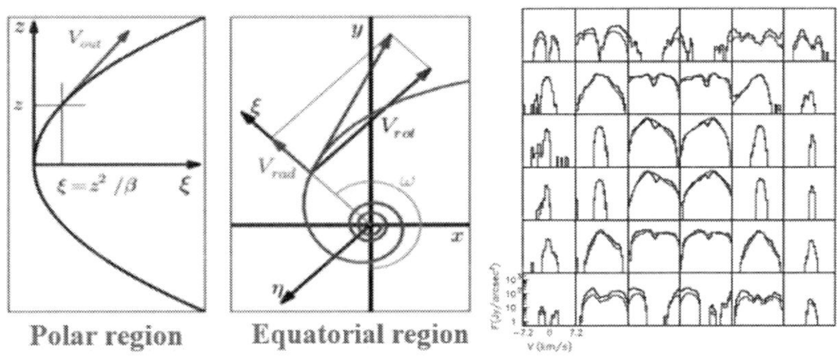

Fig. 4 Red Rectangle. Left and centre: gas velocity and trajectories. Right: Doppler velocity spectra comparing data (blue) and model (red).

The rotation and radial velocities are found equal to 1.0 km/s and 1.6 km/s respectively. Using the velocity field model and the comparison between the two line emissions we have been able to reconstruct the spatial distributions of both temperature and density, displaying a sharp separation between equatorial and polar regions (Fig. 5).

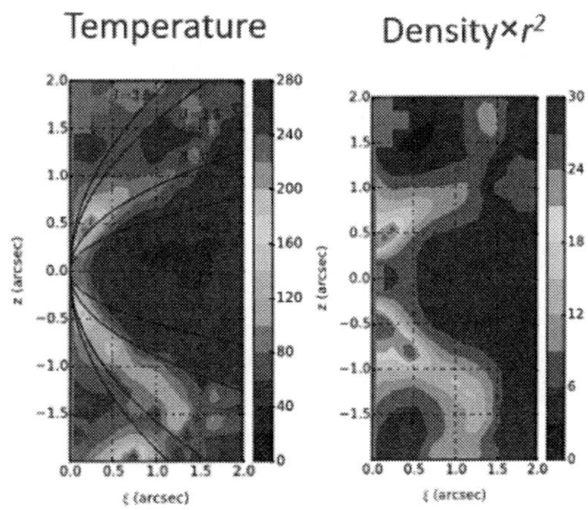

Fig. 5 Meridian distributions of temperature (left) and density multiplied by the square of the distance to the central star (right).

2.3 Young stars

Young stars form from the condensation of gas and dust in a molecular cloud. In-falling matter toward the centre of the condensed core forms rapidly an accretion disc. The cool dust accumulates in the disc plane onto which gas molecules freeze out, forming an ice mantel. Some dust grains aggregate to form larger and larger bodies, ultimately planetesimals and planets. A good summary of the current research in the field can be found in a review by Sean Andrews [20]. A topic of strong current interest is the formation of planets in such discs, called for this reason protoplanetary discs. In particular one aims at identifying clearly which physico-chemical conditions need to be fulfilled for planets to start forming. We have studied several young stars [21-23], in particular a multiple stellar system, GG Tau A, which is presented below.

GG Tau A, located at 150 pc in the Taurus molecular cloud [24,25], is considered to be an archetype of a T-Tauri – solar mass protostar– multiple system. It is a triple star made of

GG Tau Aa and a nearby binary GG Tau Ab1/Ab2 with respective separations of 35 au and 4.5 au. GG Tau A is surrounded by a circumbinary (referring to Aa and Ab) envelope of gas and dust in the form of a ring spanning from ~190 to 280 au in distance from GG Tau A and an outer disc extending up to ~800 au with a total mass ~0.15 solar masses. Using most of the available archival and new data from ALMA and NOEMA, in collaboration with experienced colleagues from University of Bordeaux, we studied the morphology, kinematics, and physical conditions of the gas and dust (distribution of temperature and density) [26,27]. The high sensitivity of the observations made it possible to explore the properties of the cavity contained inside the central ring, at distances smaller than ~140 au from the central stars, where gas is falling along two streamers onto the circumstellar discs. A preliminary study of the chemistry at stake has revealed the presence of H_2S; this was the first example of detection of the molecule in this type of protostars, made possible thanks to the relatively high mass of the GG Tau A disc [28].

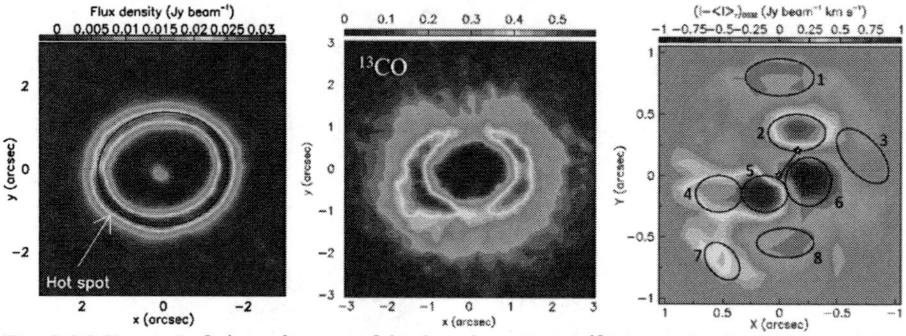

Fig. 5 GG Tau A. Left: intensity map of the dust ring. Centre: ^{13}CO gas distribution around the cavity. Right: emission in the cavity revealing the presence of gas streams.

3 Conclusions

Astrophysics, one of the most dynamic branches of modern physics, matches well the needs of a team such as ours. We have made significant contributions to the study of high redshift galaxies and a number of evolved stars and protostars.

Our experience shows that both training at home and collaboration with experienced teams abroad are mandatory for doing good research and making progress.

We are making extensive use of the archival data of the ALMA collaboration. Open access policy is a precious opportunity for astronomers, in particular in developing countries. Nowadays, most of the major observatories open their data to public: ALMA, NOEMA, and many facilities of ESO and NASA. They usually have very good supporting interface for searching, calibrating, and reducing data. Many of the data are already calibrated and ready to be used. We are all invited to use this treasure. As Pierre uses to say,

The sky belongs to all of us
We are all made of the same star dust.

Acknowledgements

All along the development of the team we have received invaluable help and support from many friends from abroad: Jim Cronin, Alan Watson, Alain Cordier, Pierre Billoir and Murat Boratav in the PAO collaboration; our present collaborators Thibaut Le Bertre, Anne Dutrey, Stéphane Guilloteau, Pierre Lesaffre and Jan Martin Winters; Antonino Zichichi and the World Association of Scientists who place in us unconditional confidence and give

https://doi.org/10.1051/epjconf/202024003001

us continuous support; Jean Tran Thanh Van and the ICISE centre in Quy Nhon who help Vietnam with its progress in developing basic research; Nguyen Quang Rieu, whom we consider as the founding father of the Vietnamese radio astronomy and gave us advice and support in the early phase of our development; Paul Ho from EAO, Young-Chol Minh from KASI, Kazuhiro Sechiguchi from NAOJ and Boonrucksar Soonthornthum from NARIT, who give us constant support and help us integrating the regional activity and contributing to its progress; in Vietnam, we had the honour to count among our friends Hoang Tuy and Viet Phuong, who are no longer with us, and today scientists such as Pham Duy Hien and Nguyen Duc Chien who follow our progress and are always prepared to give us help and advice. To all we express our deepest gratitude from the bottom of our heart.

We are deeply indebted to the ALMA partnership, whose open access policy means invaluable support and encouragement for Vietnamese astrophysics. ALMA is a partnership of ESO (representing its member states), NSF (USA), NINS (Japan), NRC (Canada), NSC/ASIAA (Taiwan), and KASI (South Korea), in cooperation with Chile. The Joint ALMA Observatory is operated by ESO, AUI/NRAO and NAOJ. Financial support from the World Laboratory, Rencontres du Viet Nam, the Odon Vallet fellowships is gratefully acknowledged. Our research is funded by Vietnam National Foundation for Science and Technology Development (NAFOSTED) under grant number 103.99-2018.325.

References

1. P.N. Dinh, N.T. Dung, B.D. Hieu, et al., Nuclear Physics B, **661**, 302 (2003)
2. P.N. Diep, P.N. Dinh, N.H. Duong, et al., Nuclear Physics B, **678**, 3 (2004)
3. R.D. Blandford, et al., Science, **245**, 824 (1989)
4. P. Saha & L.L.R. Williams, AJ, **125**, 2769 (2003)
5. D.T. Hoai, D.T. Hoai, P.Tuan-Anh, et al., RAA, **13**, 803 (2013)
6. P. Tuan-Anh, F. Boone, D.T. Hoai, et al, A&A, **552**, L12 (2013)
7. P. Tuan-Anh, D.T. Hoai, D.T. Hoai, et al., MNRAS, **467** (3), 3513-3524 (2017)
8. D.T. Hoai, L. Mathews, J.M. Winters, et al., A&A, **565**, A54 (2014)
9. P.T. Nhung, D.T. Hoai, J.M. Winters, et al., RAA, **15**, 713 (2015)
10. D.T. Hoai, P.N. Diep, N.T. Phuong, et al., RAA, **16**, 111, (2016) arXiv:1601.04834
11. D.T. Hoai, P.T. Nhung, P.N. Diep, et al., RAA, 2016, **16**, 106, arXiv:1601.01439
12. P.T. Nhung, D.T. Hoai, J.M. Winters et al., A&A, **583**, A64 (2015)
13. P.T. Nhung, P. Tuan-Anh, T. Le Bertre, et al., RAA, **19**, 43 (2019)
14. D.T. Hoai, P.T. Nhung, P. Tuan-Anh, et al., MNRAS, **484**, 1865 (2019)
15. P. Tuan-Anh, D.T. Hoai, P.T. Nhung, et al., MNRAS, **487**, 622 (2019)
16. P.N. Diep, N.T. Phuong, D.T. Hoai, et al., MNRAS **461**, 4276 (2016)
17. P.T. Nhung, D.T. Hoai, P. Tuan-Anh, et al., MNRAS, **480**, 3324 (2018)
18. D.T. Hoai, P.T. Nhung, P. Tuan-Anh, et al., Proceedings of 11th SEAAN, to be published, (2020)
19. P. Tuan-Anh, P.N. Diep, P.T. Nhung, et al., RAA, **12**, 2213 (2015)
20. S. Andrews, ARAA, arXiv: 2001.05007, to be published (2020)
21. P.T. Nhung, D.T. Hoai, P. Tuan-Anh, et al., MNRAS, **469**, 4726 (2017)
22. P. Tuan-Anh, P.T. Nhung, D.T. Hoai, et al., MNRAS, **463**, 3563 (2016)
23. P.N. Diep, D.T. Hoai, N.B. Ngoc et al., VJSTE, Vol. **61**, No. 2, p. 3-16 (2020)

24. Gaia Collaboration, A. G. A. Brown, A. Vallenari, et al., "ArXiv e-prints" arXiv: 1804.09365 (2018)
25. Gaia Collaboration, T. Prusti, J. H. J. de Bruijne, et al., A&A, **595**, A1 (2016)
26. N.T. Phuong, P.N. Diep, A. Dutrey, et al., RAA, **18**, 3, arXiv:1801.00861 (2018)
27. N.T. Phuong, A. Dutrey, P.N. Diep, et al., A&A, to be published, (2020)
28. N.T. Phuong, E. Chapillon, L Majumdar, et al., A&A, **616**, L5 (2018)

Tracing Black Hole Signatures in Several Black Hole Candidates Based on Their X-Ray Spectral Evolution

Fahmi Iman Alfarizki[1,*] and *Kiki* Vierdayanti[2]

[1]Astronomy Study Program, Faculty of Mathematics and Natural Sciences, Institut Teknologi Bandung, Indonesia
[2]Astronomy Research Division, Faculty of Mathematics and Natural Sciences, Institut Teknologi Bandung, Indonesia

Abstract. Investigation of spectral evolution of four black hole candidates was carried out by using color-color diagram as well as spectral fitting on *Swift*/XRT data. Newly found candidates, which are classified as low-mass X-ray binary system based on their transient nature, are the focus of our work. We compare their spectral evolutions to that of XTE J1752-223, a transient system and a more convincing black hole candidate whose mass has been determined from spectral-timing correlation scaling. In addition, comparison to Cygnus X-1, a well-known stellar-mass black hole, was done despite its persistent nature. The spectral fitting, by using a combination of thermal disk and non-thermal component model, results in the innermost temperature values in the range of the typical innermost temperature of black hole binary which is 0.7 – 1.5 keV. The spectral evolutions of the candidates bear a resemblance to both Cygnus X-1 and XTE J1752-223. We note that during *Swift*/XRT observations, the spectra of Cygnus X-1 and IGR J17451-3022 are mostly dominated by the non-thermal component. We conclude that the compact object of MAXI J1535-571 and MAXI J1828-249 is highly likely to be a black hole. However, the lack of data rendered conclusive result impossible for IGR J17454-2919.

Keywords: black hole binary, black hole candidates, black hole signatures, low-mass X-ray binary, spectral evolution.

1 Introduction

Direct evidence of astrophysical black hole remains difficult to obtain even after the first image of black hole's shadow , which happened to be a supermassive black hole in the center of M87 [1], has been obtained by Event Horizon Telescope (EHT) team. The difficulties are more pronounced in the case of stellar-mass black holes due to their much smaller dimension. On the other hand, their extremely compact dimension give rise to much better opportunities to study their variability. Indirect evidence, such as dynamical mass

---

*Corresponding author: fahmimn21@students.itb.ac.id

estimation in some X-ray binary systems, has been used to show the existence of stellar-mass black hole in our Galaxy and other nearby galaxies (e.g. [2-3]). However, this method cannot always be applied to X-ray binaries, especially transient systems with extremely dim optical counterpart.

Fortunately, radiation from the accretion disk surrounding a black hole in X-ray binary systems can easily be detected and it may reveal information from the innermost region near the black hole. Many well-known stellar-mass black hole candidates are discovered in X-ray binaries. It is important to note that distinguishing a stellar-mass black hole from a neutron star is not trivial due to similar relativistic effects [4]. However, since a neutron star has a solid surface while a black hole does not, we expect different accretion properties closed to the compact object. A black hole has the innermost stable circular orbit (ISCO) radius which is usually assumed as the location of the innermost radius of the accretion disk in thermal dominant state. However, extensive study of black hole binaries in our Galaxy has revealed that these sources exhibit spectral variability [5]. Done & Gierlinski (2003) have shown that spectral evolution of black hole can be distinguished from that of neutron star in X-ray binary systems. We utilized the method used in Done & Gierlinski (2003) to trace black hole signature in four newly found black hole candidates. We present our data and methods in section 2. The results of our study are presented in section 3. Section 4 is dedicated for discussion while section 5 concludes our study.

2 Data and Methods

2.1 Data Selection and Reduction

Our study focus on black hole candidates in transient systems. The list of the candidates was obtained from an on-line catalogue of stellar-mass black holes in X-ray binaries [6]. The catalogue consists of 60 stellar-mass black hole candidates and it is still growing. We chose four newly found candidates which are MAXI J1535-571, MAXI J1828-249, IGR J17451-3022, and IGR J17454-2919. The list of the candidates are presented in Table 1. Those candidates will be compared to XTE J1752-223, a more convincing candidate of black hole binary, whose mass has been determined from spectral-timing correlation scaling [7]. Also, we compared all candidates to Cyg X-1, a well-known black hole binary, despite its persistent nature.

Table 1. The comparison of amount of data, observation period, and count rates range of black hole candidates with Cyg X-1.

Objects	Number of Data	Observations Time Range (MM/YYYY)	Count Rates Range (counts/s)
Cyg X-1	113	12/2004 – 03/2019	4.09 – 914.6
XTE J1752-223	66	09/2009 – 07/2010	3.95×10^{-4} – 329.1
MAXI J1535-223	109	09/2017 – 06/2019	0.07 – 902.3
MAXI J1828-249	48	10/2013 – 04/2014	3.31×10^{-4} – 205.1
IGR J17451-3022	41	09/2012 – 05/2015	3.37×10^{-3} – 2.45
IGR J17454-2919	25	05 2007 – 09/2016	9.08×10^{-4} – 1.72

Most of data are in windowed-timing mode since the candidates are bright Galactic X-ray sources. We used Heasoft 6.26.1 for data processing and Xspec 12.10.1f for spectral fitting. We did pile-up correction for count rate greater than ~ 100 ct/s and ~ 0.5 ct/s for windowed timing mode and photon counting mode, respectively. Each data is grouped so that to have a minimum 20 counts per bin.

2.3 Spectral Fitting

We only used a combination of simple models to fit each spectra. The thermal component is modelled with multi-color disk spectrum, namely `diskbb` in Xspec, and the non-thermal component is modelled with power-law spectrum, namely `powerlaw` in Xspec. The Tübingen-Boulder ISM absorption model is also incorporated in the fitting. In our study, the chi-squared statistics can be used due to high values of total counts as well as count rate (as can be seen in Table 1) in most data. Since not all fittings result in acceptable values of reduced chi-squared, for further analysis, we only used data with $\chi^2/v < 2$.

2.4 Spectral Evolution Analysis

To get an overview of source behaviour, we used color-color and hardness-intensity diagrams. Since the count rate detected within a certain energy range depend on both the instrument response and on the observing column towards the source, we used unabsorbed flux over a given energy band to get a measure of the true source behaviour on the diagrams.

For the color-color diagram, we used three energy bands to form the soft and hard colors, 0.6 - 2.5 keV, 2.5 - 4.5 keV, and 4.5 – 10 keV. Following Reynolds & Miller (2013), the soft and hard colors are defined as the ratio of unabsorbed fluxes in the 2.5 – 4.5 keV and 0.6 – 2.5 keV bands and in the 4.5 – 10 keV and 2.5 – 4.5 keV bands, respectively. Assuming that the relative uncertainty of the unabsorbed flux is comparable to that of the count rate, we could estimate the uncertainty of the unabsorbed flux.

For the hardness – intensity diagram, the intensity was obtained from the calculation of unabsorbed fluxes in the 0.3 – 10 keV bands. As for the hardness, we defined hardness ratio as the ratio of unabsorbed fluxes in the 1.5 – 10 keV and 0.3 – 1.5 keV bands, as normally used in *Swift*/XRT data.

3 Results

3.1 Spectral fitting

Figure 1 shows the histogram of reduced chi-squared values of each candidate. We found that most of data have reduced chi-squared less than 2, although there are several data whose reduced chi-squared close to zero, which normally happens to spectra with large uncertainty. Hence, we only used data which result in reduced chi-squared between 0.5 – 2.0 for further analysis. We found that the majority of data of each candidate reveal that the innermost temperatures are in the range of black hole binary, which is between 0.5 to 1.5 keV. Despite some of them are higher or lower than that range (See Figure 2). The photon index of power-law for all candidates are very high and some of them have a wide range of photon index such as MAXI J1535-571, IGR J17451-3022, and MAXI J1828-249 as shown in Figure 3.

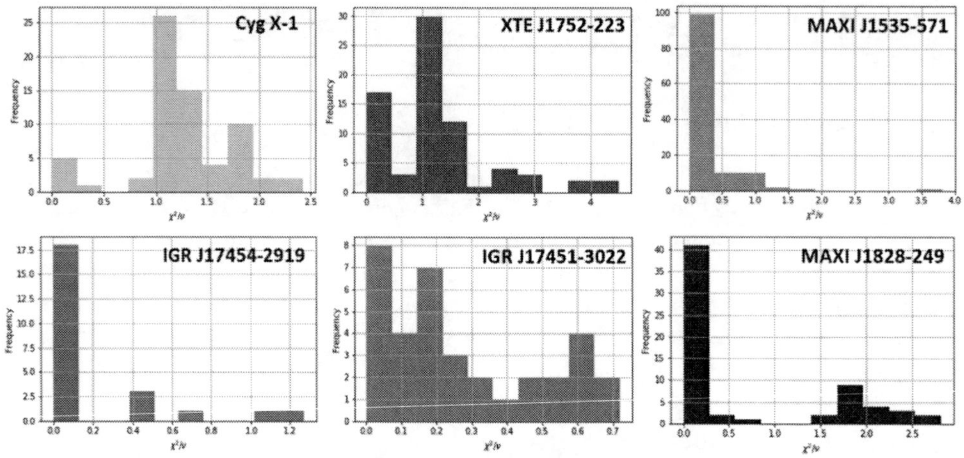

Fig. 1. Distribution of reduced chi-squared values of each candidate.

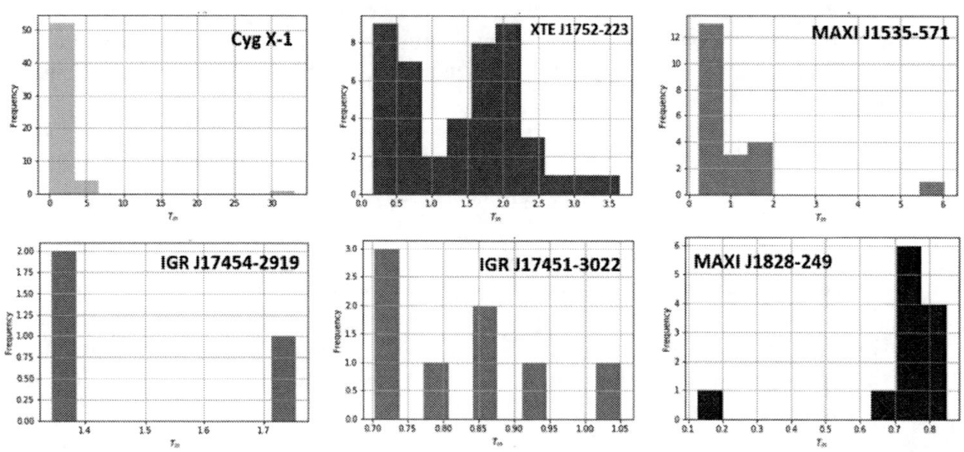

Fig. 2. Distribution of the disk innermost temperature for each candidate.

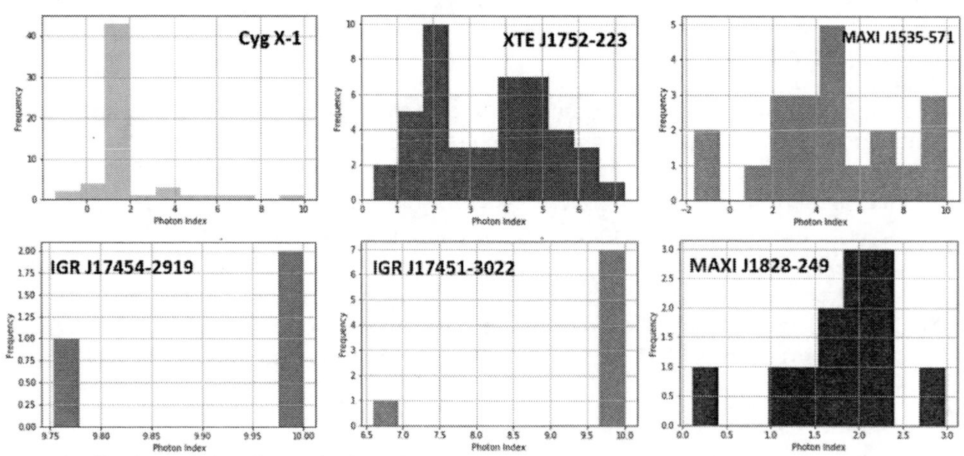

Fig. 3. Distribution of the photon index of power-law for each candidate.

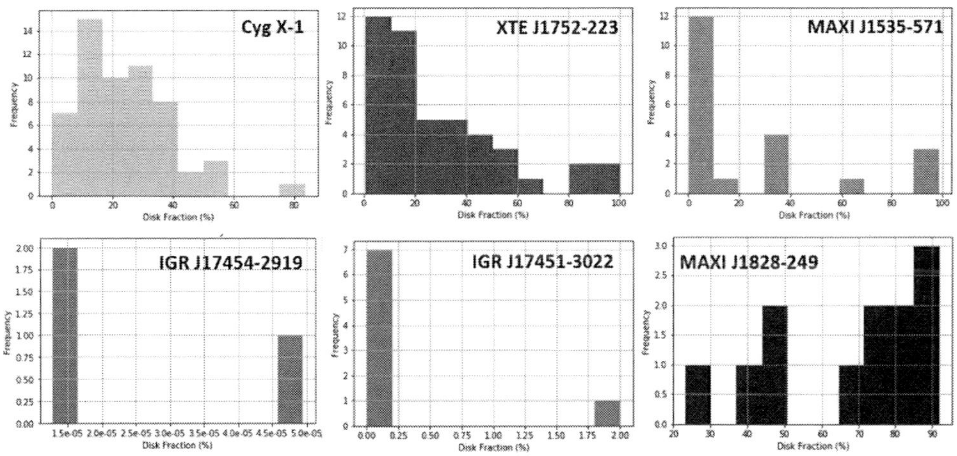

Fig. 4. Distribution of disk component flux to total flux ratio of each candidate.

We also measured the ratio between the disk component unabsorbed flux and total unabsorbed flux. Figure 4 shows that majority of data are not in thermal dominated state in which the disk contribution to the spectra must be greater than 75% [5]. Only MAXI J1828-249 has most data in thermal dominant state. From this result, we expect that Cyg X-1, IGR J17454-2919, and IGR J17451-3022 do not reveal the evidence of ISCO due to lack of data in thermal dominant state.

3.2 Spectral Evolution

Figure 5a shows the color-color diagram of Cyg X-1 and XTE J1752-223, while Figure 5b shows the color-color diagram of the candidates. The orange solid lines in both panels represents the `powerlaw` model. The blue solid lines in both panels represents the `diskbb` model. The spectral evolution of Cyg X-1 fall on the `powerlaw` model track which consistent to the fact that Cyg X-1 spends most of the time in hard state (e.g. [8,9]). The spectral evolution of XTE J1752-223 varies from hard state to ultrasoft state. Several spectra of Cyg X-1 and XTE J1752-223 occupy an area in soft color between 0.2 to 0.6 and not all candidates occupy that range. Only MAXI objects do. Others, however, are very soft. These spectral evolutions resemble those constructed by Done & Gierlinski (2003) for a sample of black holes observed by *RXTE*.

We built the light curves and hardness ratio evolution in Figure 6 to look at the transient behaviour. Cygnus X-1 reveals a persistent nature and spends most of the time in a hard state as can be confirmed by the color-color diagram. We found an outburst feature like in XTE and both MAXI objects. While in both IGR objects, the range of count rate is too narrow for a typical outburst episode of transient black hole binary. We can also see that the hardness ratio of both IGR objects are very low, implying that the thermal component predominates the spectra. Also, there is no type I burst discovered on the light curves of all candidates.

 https://doi.org/10.1051/epjconf/202024004001

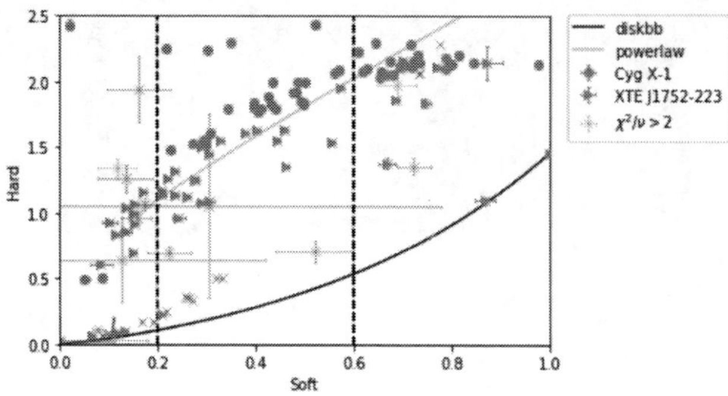

(a)

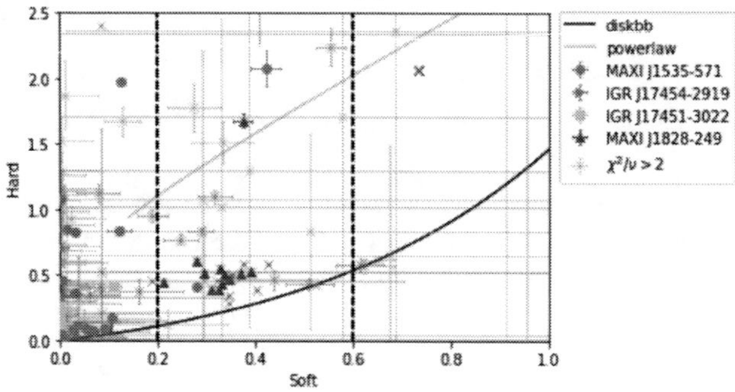

(b)

Fig. 5. Color-color diagram of Cyg X-1 and XTE J1752-223 (a) and all candidates (b). The dashed black lines are soft color which correspond to 0.7 – 1.5 keV of the disk innermost temperature.

In the hardness-intensity diagram as shown in Figure 7, we can trace the black hole spectral evolution. Cyg X-1 and XTE 1752-223 show Q-shaped pattern on the spectral evolution track. These consistent to black hole binary evolution. When the luminosity is high, then soft component is dominant, and vice versa. Resemblance to the spectral evolution track to CygX-1 and XTE 1752-223 can only be seen in MAXI 1535-571. The spectral evolution track of the rest of the candidates is inconclusive due to the lack of data.

4 Discussion

The area in soft color between 0.2 to 0.6 in the color-color diagrams in Figure 5b can not be occupied by neutron stars because the neutron stars have a additional flux from their surface [8]. Hence, MAXI J1828-249 behavior in color-color diagram is in a good agreement with that of black hole. For the rest of the candidates, however, we cannot, based on the color-color diagram, see their resemblance to that of black hole. Their extremely soft color indicate resamblence to that of super-soft source which are known to host a white dwarf.

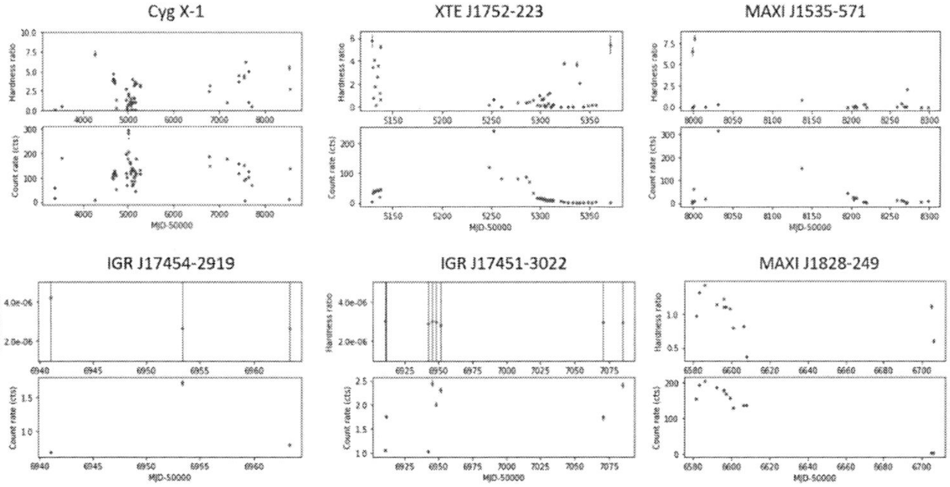

Fig. 6. The upper panel is the hardness ratio evolution and the lower panel is the light curve in terms of count rate.

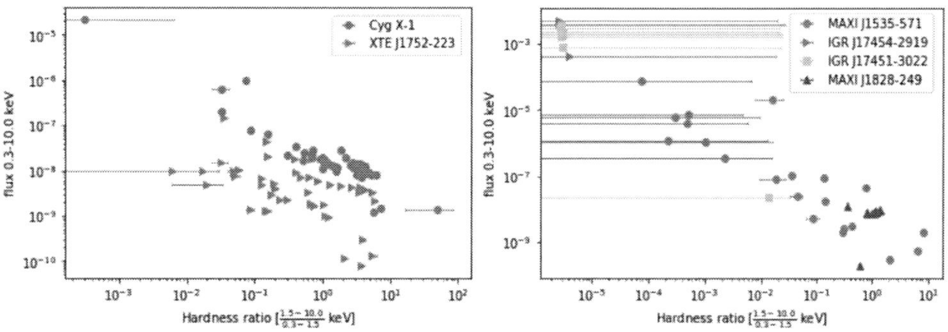

Fig. 7. Hardness-intensity diagrams of Cyg X-1 and XTE J1752-23 (left-hand panel) and the candidates (right-hand panel).

Figure 7 reveal that MAXI J1535-571 clearly follow the spectral evolution tracks of Cyg X-1 and XTE J1752-223. Although MAXI J1828-249, IGR J17454-2919, and IGR J17451-3022 do not show full spectral evolution tracks, we cannot rule out the possibility that MAXI J1828-249 and both IGR objects follow the spectral evolution track of Cyg X-1 and XTE J1752-223. Another evidence for the black hole event horizon can be obtained from the values of the innermost radius of accretion disk. However, it is necessary to check the relation between the innermost disk temperature and the luminosity to make sure that the spectra are in thermal dominated state.

In thermal dominated state, the innermost radius of the accretion disk can be assumed to reach the ISCO radius. Moreover, we expect that the disk luminosity and the innermost temperature of the disk obey $L \propto T^4$. Since we do not have the distance parameter, we measured the disk luminosity in term L/d^2. Further, we also expect that the innermost radii of accretion disk remain constant and can be assumed to coincide with the ISCO radius.

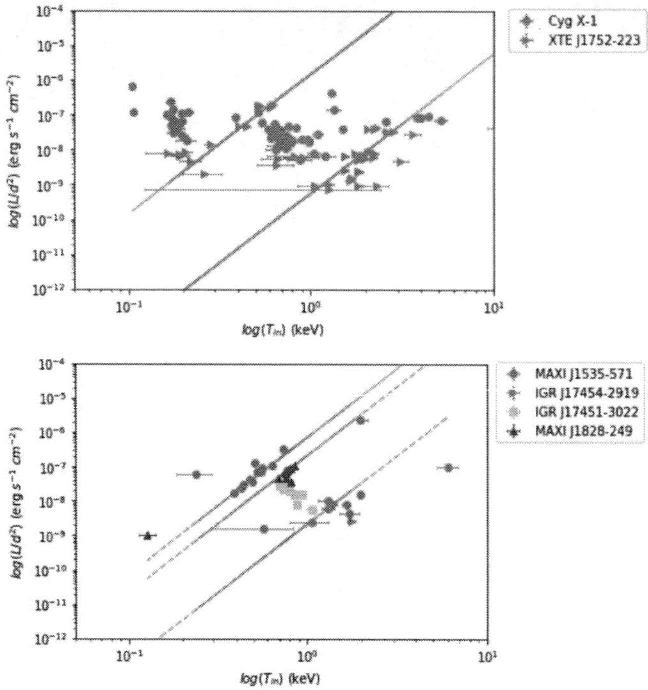

Fig. 8. The innermost disk temperature and luminosity relations. The lines are the L/d^2 as a function of T^4.

Figure 8 shows the relation between the innermost disk temperature and luminosity of Cyg X-1 and XTE J1752-223 (upper panel) and all candidates (lower panel). Some spectra of XTE J1752-223 obey $L \propto T^4$ relation and we can expect that this black hole shows the evidence of ISCO and thus the evidence of black hole event horizon. Cyg X-1 clearly does not obey $L \propto T^4$ relation since most of time it is in the hard state. The persistent nature of Cyg X-1 can be seen in this diagram in which the values of L/d^2 remain high and constant. Both MAXI objects obey $L \propto T^4$ relation in which we can expect that the spectra are in thermal dominated state. Moreover, MAXI objects also reveal the existence of ISCO in the luminosity-radius diagram (Figure 9). IGR J17451-3022 does not obey the relations and this result consistent to the fact that non-thermal component is dominant in all spectra of IGR J17451-3022 as shown in Figure 4. Some data of IGR J17454-2919 fall into the line of $L{\sim}T^4$ relation, but we cannot conclude that the spectra are in thermal dominated state.

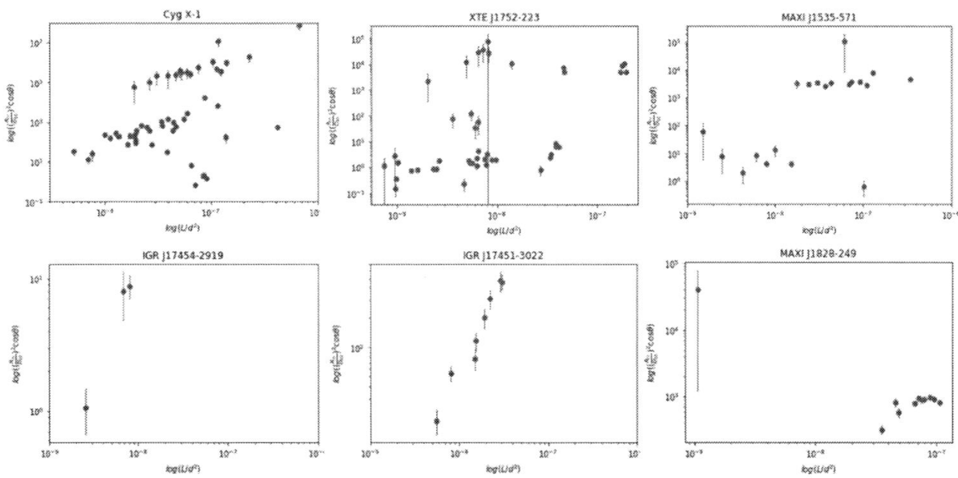

Fig. 9. The luminosity and the innermost radius relations. Since we do not have the distance and the inclination of the disk, we used normalization of diskbb which is correlated to $(R_{in}/D_{10})^2cos\theta$.

As we expected in previous discussion, XTE J1752-223, MAXI J1535-571, and MAXI J1828-249 reveal the existence of the ISCO in some spectra in which the innermost radius of accretion disk remain constant (See Figure 9). However, we did not discover the existence of the ISCO in Cyg X-1 and IGR J17451-3022. This result is consistent to the fact that Cyg X-1 and IGR J17451-3022 are mostly in hard state. Unfortunately, we cannot find any evidence of black hole for IGR J17454-2919, due mostly to the lack of data.

5 Conclusion

Spectral evolution of the black hole candidates has been investigated and compared to those of Cyg X-1 and XTE J1752-223. The spectral evolution of MAXI J1535-571 is very similar to those of Cyg X-1 and XTE J1752-223. Although MAXI J1828-249, IGR J17454-2919, and IGR J17451-3022 do not show full spectral evolution tracks, we cannot rule out the possibility that MAXI J1828-249 and both IGR objects follow the spectral evolution track of Cyg X-1 and XTE J1752-223. MAXI J1535-571 and MAXI J1828-249 reveal an evidence of ISCO radius which is the most prominent black hole signatures. Innermost disk radii of IGR J17451-3022 do not remain constant because the spectral evolutions are not in thermal state. Nevertheless, IGR J17451-3022 reveals several evidences and it is more likely to be a black hole. Because lack of data, IGR J17454-2919 does not reveal the evidence of ISCO black hole. However, it reveals some signatures of black holes and must be further examined.

References

1. The EHT Collaboration et al., Astrophys. J. Lett., *First M87 Event Horizon Telescope Results. I. The Shadow of the Supermassive Black Hole*, **875**, L1 (2019)
2. A. H. Prestwich, R. Kilgard, P. A. Crowther, S. Carpano, A. M. T. Pollock, A. Zezas, S. H. Saar, T. P. Roberts and M. J. Ward, Astrophys. J. Lett., *The Orbital Period of the Wolf-Rayet Binary IC 10 X-1: Dynamic Evidence that the Compact Object Is a Black Hole*, **669**, L21-L24 (2007)

3. D. Mata Sánchez, T. Muñoz-Darias, J. Casares, J. M. Corral-Santana, and T. Shahbaz, Mon. Not. R. Astron. Soc., *Swift J1357.2-0933: A Massive Black Hole in the Galactic Thick Disc*, **454**, 2199-2204 (2015)
4. N. Ramesh, A&G, *Evidence of the Black Hole Event Horizon*, **44**, 6.22-6.25 (2003)
5. R. A. Remillard and J. E. McClintock, Annu. Rev. Astro. Astrophys., *X-Ray Properties of Black-Hole Binaries*, **44**, 49-92 (2006)
6. J. M. Corral-Santana, J. Casares, T. Muñoz-Darias, F. E. Bauer, I. G. Martínez-Pais, and D. M. Russell, Astron. Astrophys, *BlackCAT: A Catalogue of Stellar-mass Black Holes in X-ray Transients*, **587**, A61 (2016)
7. N. Shaposhnikov, C. Markwardt, J. Swank and H. Krimm, Astrophys. J., *Discovery and Monitoring of a New Black Hole Candidate XTE J1752-223 with RXTE: Rms Spectrum Evolution, Black Hole Mass, and the Source Distance*, **723**, 1817-1824 (2010)
8. C. Done and M. Gierlinski, Mon. Not. R. Astron. Soc., *Observing the Effects of Event Horizon in Black Holes*, **342**, 1041-1055 (2003)
9. M. T. Reynolds and J. M. Miller, Astrophys. J., *A Swift Survey of Accretion onto Stellar-Mass Black Holes*, **769**, 16 (2013)

Determination of the local dark matter density using K-dwarfs from Gaia DR2

M. Dafa Wardana[1,*], *Hesti* Wulandari[2,**] *Sulistiyowati*[3,***], and *Akbar* H. Khatami[1,****]

[1]Astronomy Study Program, Institut Teknologi Bandung, West Java, Indonesia
[2]Astronomy Research Division, Institut Teknologi Bandung, West Java, Indonesia
[3]SkyLabs, Depok, West Java, Indonesia

Abstract. Local dark matter density, ρ_{dm}, is one of the crucial astrophysical inputs for the estimation of detection rates in dark matter direct search experiments. Knowing the value also helps us to investigate the shape of the Galactic dark halo, which is of importance for indirect dark matter searches, as well as for various studies in astrophysics and cosmology. In this work, we performed kinematics study of stars in the solar neighborhood to determine the local dark matter density. As tracers we used 95,543 K-dwarfs from Gaia DR2 inside a heliocentric cylinder with a radius of 150 pc and height 200 pc above and below the Galactic mid plane. Their positions and motions were analyzed, assuming that the Galaxy is axisymmetric and the tracers are in dynamical equilibrium. We applied Jeans and Poisson equations to relate the observed quantities, i.e. vertical position and velocity, with the local dark matter density. The tilt term in the Jeans equation is considered to be small and is therefore neglected. Galactic disk is modelled to consist of a single exponential stellar disk, a thin gas layer, and dark matter whose density is constant within the volume considered. Marginalization for the free parameters was performed with Bayesian theorem using Markov Chain Monte Carlo (MCMC) method. We find that $\rho_{dm} = 0.0116 \pm 0.0012\ M_\odot/pc^3$ or $\rho_{dm} = 0.439 \pm 0.046\ GeV/cm^3$, in agreement within the range of uncertainty with results of several previous studies.

1 Introduction

Efforts to understand the properties of dark matter are constantly being made through both indirect and direct experiments. The later detect interactions between dark matter particles and nuclei in the detector materials, which are usually assembled underground. For direct experiments, local dark matter density, ρ_{dm}, is an important parameter for estimation of the detection rate of dark matter particles. ρ_{dm} is defined as the value of dark matter density in the solar neighborhood. Although there are no tight restrictions set by the term "local", ρ_{dm} is commonly determined by examining a small volume around the Sun, typically within the height of a few hundred parsecs up to $\sim 10^3$ pc from the Galactic mid plane [1].

*e-mail: dafaward@gmail.com
**e-mail: hesti@as.itb.ac.id
***e-mail: sulis.astro08@gmail.com
****e-mail: akbar.h.khatami@gmail.com

Aside from its importance to estimate the detection rate of dark matter particles, it is also intriguing to obtain the value of ρ_{dm}, because knowing it can help revealing the actual shape of the Galactic dark halo – which has been assumed to have a spherical symmetry. This is done by comparing ρ_{dm} with the value obtained from global dark matter density profile through decomposition of Milky Way's rotation curve [2, 3]. If the value of ρ_{dm} is smaller, then the Galaxy has a prolate-shaped halo. A higher value of ρ_{dm} indicates that Milky Way's halo is oblate with a possibility of a dark-disk structure.

The value of ρ_{dm} from determinations in the past few years lies within the range of 10^{-3} $M_\odot/pc^3$ and 0.05 $M_\odot/pc^3$ [4–9]. Methods to determine local dark matter density are constantly being updated along with the increase of the quality and amount of observational data, such as obtained by Gaia, a mission initiated by the European Space Agency (ESA) with astrometry as the main focus. In particular Gaia measures positions and velocities of stars in the Galaxy [10]. Data obtained by Gaia has a precision up to miliarcseconds for position and proper motion measurements, and a range of magnitude of $3 \leq G \leq 21$ for photometric measurement. Gaia's photometrics G passband covers a wavelength range between 330 nm and 1050 nm. At the moment, the most recent data available is Gaia DR2, published on 25th April 2018 [11].

2 Method

One of the possible approaches to determine the local dark matter density is by observing the motion of stars near the Sun as tracers. Stars move based on their perceived gravitational potential, while gravitational potential is generated by mass, both luminous or dark matter. If we are able to model the distribution of baryonic matter, then the difference between the total density and the density of baryonic matter informs the density of non-baryonic dark matter.

The populations of stars selected as tracers are, where possible, the ones in dynamical equilibrium. Dynamical equilibrium is important as it can simplify the collisionless Boltzmann equation. Generally, early-type stars of class O and B are never in dynamic equilibrium due to their short lifetime. Several similar works by Garbari et al. (2012)[4], Xia et al. (2016)[7], and Sivertsson et al. (2018)[8], for instance, with the consideration that a long lifetime will produce a larger chance for a star population to reach dynamical equilibrium, selected longer-lived stars as tracers, rather than early-type stars.

2.1 Jeans equation

As we can assume the stars in the Galaxy as a collisionless system, to study its dynamics, we can use the collisionless Boltzmann equation

$$\frac{df}{dt} = \frac{\partial f}{\partial t} + \nabla_x f \cdot v - \nabla_v f \cdot \nabla\Phi = 0, \tag{1}$$

where f is a six dimensional phase space distribution (three dimensions for space and three dimensions for velocity) of stars in the Galaxy, v is star velocity, and Φ is gravitational potential.

The collisionless Boltzmann equation fulfills $df/dt = 0$. This means that the phase space distribution for a particle does not change with respect to time. Hence, if we follow a star in its orbit, the six-dimensional phase space density surrounding the star is constant. For a system in dynamical equilibrium, the term $\partial f/\partial t$ is 0. This study assumes that the populations of stars used as tracers are in dynamical equilibrium.

If the collisionless Boltzmann equation is integrated with respect to velocity, the Jeans equation will be acquired. In a cylindrical coordinate system, the Jeans equation can be expressed as

$$-\nu\frac{d\Phi}{dz} = \frac{1}{R}\frac{\partial}{\partial R}(R\nu\sigma_{Rz}) + \frac{1}{R}\frac{\partial}{\partial\phi}(\nu\sigma_{\phi z}) + \frac{d}{dz}(\nu\sigma_z^2) \tag{2}$$

where ν is the tracer number density, and σ is the tracer vertical velocity dispersion. Respectively, $-\frac{d\Phi}{dz} = K_z$, $\frac{1}{R}\frac{\partial}{\partial R}(R\nu\sigma_{Rz}) = T(z)$, and $\frac{1}{R}\frac{\partial}{\partial\phi}(\nu\sigma_{\phi z}) = A(z)$ are called vertical acceleration, tilt term, and axial term. We assume that the Milky Way is axisymmetric, meaning the axial term in the Jeans equation can be neglected [4, 7].

In the Jeans equation, the motion of radial and vertical components are only coupled through the first term in the right hand side called the tilt term. Read (2014)[1] explains how the tilt term can be neglected with respect to other terms for the region in proximity to the Galactic plane, reducing Jeans equation into only in the z direction

$$\frac{d}{dz}[\nu(z)\sigma_z^2(z)] = -\nu\frac{d\Phi(z)}{dz}. \tag{3}$$

2.2 Poisson Equation

The relation between gravitational potential and matter density is explained through the Poisson equation

$$\nabla^2\Phi = 4\pi G\rho_{tot}, \tag{4}$$

where ρ_{tot} being the total matter density. If Equation 4 is expressed in a cylindrical coordinate system, it reads

$$\frac{\partial^2\Phi}{\partial z^2} + \frac{1}{R}\frac{\partial}{\partial R}R\frac{\partial\Phi}{\partial R} + \frac{1}{R^2}\frac{\partial^2\Phi}{\partial\phi^2} = 4\pi G\rho_{tot}. \tag{5}$$

Under the assumption that the Galaxy is axisymmetric, the potential derivative in the ϕ direction can be neglected, hence

$$\frac{1}{4\pi G}\frac{\partial^2\Phi}{\partial z^2} + \frac{1}{4\pi GR}\frac{\partial}{\partial R}(R\frac{\partial\Phi}{\partial R}) = \rho_{tot}. \tag{6}$$

The term $\frac{1}{4\pi GR}\frac{\partial}{\partial R}(R\frac{\partial\Phi}{\partial R}) = \frac{1}{4\pi GR}\frac{\partial V_C^2}{\partial R}$ is known as the rotation curve term. Observations show that for the region in proximity to the Sun, the rotation curve of the Galaxy tends to be flat, resulting in a small value of $\partial V_C^2/\partial R$. The value of this term can be calculated using Oort's constants A and B [12],

$$\frac{1}{4\pi GR}\frac{\partial V_C^2}{\partial R} = \frac{B^2 - A^2}{2\pi G}, \tag{7}$$

with A and B as Oort's constants.

Bovy (2017)[13] gives the value of $A = 15.3 \pm 0.4$ km s^{-1} kpc^{-1}, and $B = -11.9 \pm 0.4$ km s^{-1} kpc^{-1}, which we used in our work. Substituting those values, we have the value of rotation curve term to be around $\sim (-10^{-22})$. Therefore, that term shall be neglected with respect to the other terms. Thus, Poisson equation is reduced to only in the z direction,

$$\frac{d^2\Phi(z)}{dz^2} = 4\pi G\rho_{total}(z). \tag{8}$$

The total matter density, ρ_{total}, is assumed to consist of stars, gas, and dark matter.

$$\rho_{total}(z) = \rho_*(z) + \rho_{gas}(z) + \rho_{dm}(z). \tag{9}$$

2.3 Vertical Acceleration Modelling

Vertical acceleration is the first derivative of potential with respect to the vertical distance,

$$K_z(z) \equiv -\frac{d\Phi(z)}{dz}. \tag{10}$$

Substituting to Equation 3, we will obtain

$$\frac{d}{dz}[\nu(z)\sigma_z^2(z)] = \nu(z)K_z(z). \tag{11}$$

To model vertical acceleration, we follow Xia et al. (2016)[7]. Several additional assumptions need to be taken:

- the vertical distribution of stellar mass follows an exponential profile, $\rho_*(z) = \rho_{*,0}\exp(-z/h)$.
- gas is distributed in a negligible thickness. Hence, the surface density of the gas, Σ_{gas}, is assumed to be independent of z. In this study, we used the value given by McKee et al. (2015)[14] for the gas surface density.
- the dark matter density is constant within the considered volume, $\rho_{dm}(z) = \rho_{dm}$.

Using those assumptions, the vertical distribution of matter can be expressed as

$$\rho_{total}(z) = \rho_{*,0}\exp(-z/h_*) + \Sigma_{gas,0}/\delta(z) + \rho_{dm}. \tag{12}$$

The expression for vertical acceleration can be acquired by integrating Poisson equation

$$K_z(z) = -\int_0^z 4\pi G\rho_{total}(z')dz', \tag{13}$$

that gives

$$K_z(z) = -2\pi G\left\{\Sigma_*\left[1 - \exp\left(-\frac{z}{h_*}\right)\right] + \Sigma_{gas} + 2\rho_{dm}z\right\}, \tag{14}$$

where Σ_* is the surface mass density of the stellar disk.

The vertical velocity dispersion can be obtained analytically by integrating Equation 11

$$\int_{z_0}^{z} dz'\frac{d}{dz'}[\nu(z)\sigma_z^2(z)] = \int_{z_0}^{z} dz'\nu(z')K_z(z'). \tag{15}$$

To compute the integral, the tracer number density is modeled with an exponential profile

$$\nu(z) = \nu(z_0)\exp\left(\frac{-z}{h_t}\right), \tag{16}$$

thus, Equation 15 becomes

$$\sigma_z^2(z) = \frac{1}{\nu(z)}[\int_{z_0}^{z}\nu(z')K_z(z')dz' - \nu(z_0)\sigma_z^2(z_0)], \tag{17}$$

that gives

$$\sigma_z^2(z) = f(z) + \frac{\nu(z_0)\sigma_z^2(z_0) - \nu(z_0)f(z_0)}{\nu(z_0)\exp\left(-\frac{z}{h_t}\right)}, \tag{18}$$

where

$$f(z) = 2\pi G h_t\left\{\Sigma_*\left[1 - \frac{h_*}{h_t + h_*}\exp\left(-\frac{z}{h_*}\right)\right] + \Sigma_{gas} + 2\rho_{dm}(z + h_t)\right\}. \tag{19}$$

Equation 18 is the main equation to be solved in this study to find the local dark matter density ρ_{dm}, based on the observation of tracers' number density $\nu(z)$, and the tracers' vertical velocity dispersion σ_z.

3 Data

The considered volume in this study is a cylinder with a radius of 150 pc (0.9% of the radius of the Milky Way disk) and a half-height of z = 200 pc with the Sun on the cylinder's axis and the Galactic plane dividing the cylinder into two equal parts. There is no strict rules for choosing the considered volume size and shape. However, some considerations need to be taken into account. First, the considered space should be relatively small in comparison to the Milky Way's disk, which means that the Galaxy rotation velocity in the considered volume is relatively constant. Second, the considered volume has to be relatively small compared to the dark matter scale-length, which is 19 kpc [1], so that the dark matter density in the considered volume can be assumed to be constant. Finally, a cylindrical shaped was chosen to ease the data processing steps, which involve two equations which are functions of z.

Stellar population that is used as tracers must satisfy these following criteria:

1. the population must be old enough that it is possible for it to have reached dynamical equilibrium. Hence, collisionless Boltzmann equation can be simplified by omitting $\partial f/\partial t$.

2. the population must contain a sufficiently large number of stars to represent statistically good sample size.

3. the members of the population must be spread throughout the considered volume.

Unfortunately, Gaia DR2 has several limitations that narrow the options for tracers. The shortfall of Gaia relevant to our case are: the range of effective temperatures for stars with T_{eff} data lies within $3,000K < T_{eff} < 10,000K$ and only stars within the magnitude range of $4 < G < 13$ and effective temperature range of $3,550 < T_{eff} < 6,900K$ have radial velocity data. Realizing Gaia DR2 restrictions and concerning our criteria for tracers, we select K-dwarfs as tracers in this study. Our selection process gives 95,543 K-dwarfs with positional data, so that they can be used in fitting of tracer number density. Among them, 46,466 having data of position and velocity and can be used to calculate vertical velocity dispersion.

3.1 The likelihood

The free parameter marginalization was performed based on Bayesian theorem using Markov Chain Monte Carlo (MCMC) method. Bayesian theorem can be expressed as

$$P(\theta \mid D, M) = \frac{P(D \mid \theta, M)P(\theta \mid M)}{P(D \mid M)}, \tag{20}$$

where
θ = parameters
D = data
M = model
$P(\theta \mid D, M)$ = posterior
$P(D \mid \theta, M)$ = likelihood function
$P(\theta \mid M)$ = prior
$P(D \mid M)$ = normalization factor.
The normalization factor can be written as

$$P(D \mid M) = \int P(D \mid \theta, M)P(\theta \mid M)d\theta. \tag{21}$$

However, the normalization factor does not affect the calculation of posterior. Thus, Equation 20 can be expressed in proportional form as

$$P(\theta \mid D, M) \propto P(D \mid \theta, M)P(\theta \mid M). \tag{22}$$

The likelihood function used in this study is,

$$L(\theta) = \exp\left(-\frac{\chi^2}{2}\right), \tag{23}$$

$$\chi^2 = \sum_j \frac{[\sigma_{data,j}(z) - \sigma_{model,j}(z)]^2}{\epsilon^2_{\sigma_j(z)}}, \tag{24}$$

where L is the likelihood, $\sigma_{data,j}(z)$ is the vertical velocity dispersions as a function of height obtained from observational data, $\sigma_{model,j}(z)$ is the vertical velocity dispersion as a function of height based on the model explained in section 2, and $\epsilon_{\sigma_j(z)}$ is the vertical velocity dispersion errors from observational data.

3.2 Tracer Number Density Fitting

The considered cylindrical volume was divided in 10 pc bins in the z direction. The tracer number density was obtained by calculating the number of tracers in a bin divided by with the volume of the bin. The Galaxy was assumed to be symmetrical to the Galactic plane.

The fitting error of $\nu(z_0)$ and h_t was approximated by equations,

$$\epsilon_{\nu(z_0)} = \epsilon_{b_0}\nu(z_0), \tag{25}$$

and

$$\epsilon_{h_t} = \epsilon_{b_1}h_t, \tag{26}$$

where

$$\epsilon^2_{b_0} = \frac{\sum(z_i^2)}{N\sum(z_i - \bar{z})^2}\epsilon^2, \tag{27}$$

$$\epsilon^2_{b_1} = \frac{\epsilon^2}{\sum(z_i - \bar{z})^2}, \tag{28}$$

$$\epsilon^2 = \frac{1}{N-m}\sum\{\ln[\nu_{data}(z)] - \ln[\nu_{model}(z)]\}^2, \tag{29}$$

where N, $\bar{z}$, $\epsilon_{\nu(z_0)}$, ϵ_{h_t}, and m are the number of tracers in a bin, the average height of tracers in a bin, the tracer number density error on the Galactic plane, the tracer altitude scale error, and the number of parameters, respectively. We have two parameters fitting here, namely $\ln[\nu(z_0)]$ and h_t. Meanwhile, ϵ_{b_0} and ϵ_{b_1} are errors of parameters b_0 and b_1. From the fitting, the value of tracer number density on Galaxy plane, $\nu(z_0) = 0.0024 \pm 0.0001\ pc^{-3}$, and tracer altitude scale, $h_t = 427.25 \pm 1.54$ pc, were obtained.

3.3 Tracer Vertical Velocity Dispersion Calculation

Tracer vertical velocity, v_z, was obtained using data of equatorial coordinates, proper motion, parallax, and radial velocity. Velocity transformations from $\mu_{\alpha*}, \mu_\delta$ and v_r into v_x, v_y, and v_z were done using library *astropy.coordinates*. Then, v_z error was calculated respecting error propagation principle using these following equations,

$$\epsilon^2_{v_z} = [\cos(b)]^2\epsilon^2_{v_{t,b}} + [\sin(b)]^2\epsilon^2_{v_r} + [v_r\cos(b) - v_{t,b}\sin(b)]^2\epsilon^2_b, \tag{30}$$

$$\epsilon^2_{v_{t,b}} = \left(-4,74\frac{\mu_b}{\varpi^2}\right)^2 \epsilon^2_{\varpi} + \left(\frac{4,74}{\varpi}\right)^2 \epsilon^2_{\mu_b}, \tag{31}$$

$$\epsilon^2_{\mu_b} = A^2\epsilon^2_{\mu_\delta} + B^2\epsilon^2_{\mu_{\alpha*}} + C^2\epsilon^2_{\delta} + D^2\epsilon^2_{\alpha} + E^2\epsilon^2_{b}, \tag{32}$$

$$\epsilon^2_b = [\sin(\delta_G)\cos(\delta) + \cos(\delta_G)\sin(\delta)\sin(\alpha)]^2\epsilon^2_\delta + [-\cos(\delta_G)\cos(\delta)\cos(\alpha)]^2\epsilon^2_\alpha, \tag{33}$$

where

$$A = \frac{\sin(\delta_G)\cos(\delta)}{\cos(b)} - \frac{\cos(\delta_G)\sin(\delta)\cos(\alpha-\alpha_G)}{\cos(b)}, \tag{34}$$

$$B = \frac{\cos(\delta_G)\sin(\alpha-\alpha_G)}{\cos(b)}, \tag{35}$$

$$C = -\frac{\mu_\delta\sin(\delta_G)\sin(\delta)}{\cos(b)} - \frac{\mu_{delta}\cos(\delta_G)\cos(\delta)\cos(\alpha-\alpha_G)}{\cos(b)}, \tag{36}$$

$$D = \frac{\mu_\delta}{\cos(b)}\sin(\delta_G)\sin(\alpha-\alpha_G) - \frac{\mu_{\alpha*}}{\cos(b)}\cos(\delta_G)\cos(\delta)\cos(\alpha-\alpha_G), \tag{37}$$

$$E = E_1 - E_2 - E_3, \tag{38}$$

$$E_1 = \mu_\delta\tan(b)\sec(b)\sin(\delta_G)\cos(b), \tag{39}$$

$$E_2 = \mu_\delta\tan(b)\sec(b)\cos(\delta_G)\sin(\delta)\cos(\alpha-\alpha_G), \tag{40}$$

$$E_3 = \mu_{\alpha*}\tan(b)\sec(b)\cos(\delta_G)\sin(\alpha-\alpha_G). \tag{41}$$

In the equations above, μ_b is proper motion in the galactic latitude direction, $v_{t,b}$ is tangential velocity in the galactic latitude direction, $\delta_G = 27.12825°$ is the declination of the north galactic pole, and $\alpha_G = 192.85948°$ is the right ascension of the North Galactic Pole [15].

The tracer vertical velocity dispersion, $\sigma_{data}(z)$, for every bin was calculated using equation

$$\sigma_{data} = \sqrt{\frac{\sum_i(v_{z,i}-\bar{v})^2}{N-1}}, \tag{42}$$

where $\bar{v}$ is the average vertical velocity for each bin. Velocity dispersion error for each was calculated following steps described by Pryor & Meylan (1993)[16].

Data processing was then continued by running the MCMC program, using velocity dispersion, velocity dispersion error, tracer height-scale, and tracer number density previously obtained (see sub section 3.2). The solution to Equation 18 was determined using MCMC program. Local dark matter density (ρ_{dm}), stellar surface density (Σ_*), stellar scale-height (h_*), and tracer scale-height (h_t) were set as free parameters, whose values were acquired based on the likelihood distribution within a certain range of priors.

3.4 Priors for the Free Parameters

The range of priors for the MCMC method were selected based on preceding knowledge available from literature. The prior for h_* refers to the typical scale-height value of the thin disk, where a range of 200 pc to 400 pc was used. The typical scale-height value of the thin disk was taken as a benchmark, because the height range considered in this study was between 0 pc to 200 pc from the Galactic plane and within this range the stellar thin disks should still dominate stellar the distribution [17]. The range allowed for the Σ_* sample retrieval was between 20 $M_\odot/pc^2$ to 50 $M_\odot/pc^2$, refering to the value obtained by Xia et al. (2016)[7] and Zhang et al. (2013)[5], which is 42 ± 5 $M_\odot/pc^2$ dan 39.1 ± 7.1 $M_\odot/pc^2$. The prior for ρ_{dm} were taken based on previous works such as by Garbari et al. (2012)[4], Zhang et al. (2013)[5], Bienaymé et al. (2014)[6], Xia et al. (2016)[7], Buch et al. (2018)[9], and Sivertsson et al.

(2018)[8]. We decided to give it a range of 0 M$_\odot$/pc^3 to 0.1 M$_\odot$/pc^3. Even though the value of h_t has been obtained from number density fitting, it was still set as a free parameter, albeit with a relatively narrower prior value span. This was done to accommodate errors from the fitting results. The prior range used was 422.25 pc – 432.25 pc.

4 Result and Comparison with other Determinations

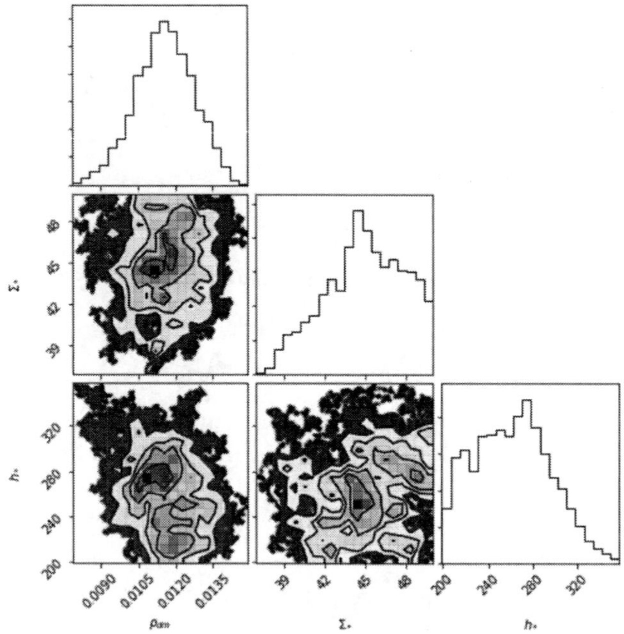

Figure 1. The marginalized posterior for the three free parameters. The units for ρ_{dm}, Σ_*, and h_* are respectively M$_\odot$/pc^3, M$_\odot$/pc^2, and pc.

The value of ρ_{dm} obtained in this study is 0.0116 ± 0.0012 M$_\odot$/pc^3 or ρ_{dm} = 0.439 ± 0.046 GeV/cm^3. As shown by PDFs of the free parameters on Figure 1, Σ_* and h_* are not tightly constrained, indicating that the model we used is insensitive to both quantities. This is mainly because the terms from the main equation (Equation 18) containing Σ_* and h_* are more complex than the linear term containing ρ_{dm}.

A number of local dark matter density determinations carried out by several researchers in the past few years are listed in Table 1 and plotted in Figure 2. As shown in Table 1, the value we obtained is smaller than the results obtained by Garbari et al. (2012)[4], Bienaymé et al. (2014)[6], Xia et al. (2016)[7], and Sivertsson et al. (2018)[8], but is higher than the results obtained by Zhang et al. (2013)[5]. Meanwhile, Buch et al. (2018)[9], who also used Gaia DR2 to determine the local dark matter density (but using different tracers from this study), also obtained higher results than is obtained in this study. But within the error ranges, the result of this study agrees with that of Garbari et al. (2012)[4], Xia et al. (2016)[7], Sivertsson et al. (2018)[8], and Buch et al. (2018)[9] for the G-dwarfs tracer.

The difference in the value of ρ_{dm} we obtained with the value of ρ_{dm} in the aforementioned studies can be attributed to several factors as follows:

 https://doi.org/10.1051/epjconf/202024004002

Table 1. Several determinations of the local dark matter density for the recent years.

Name (year)	Tracer	ρ_{dm} ($M_\odot/pc^3$)
Garbari et al. (2012)	K-dwarfs	$0.022^{+0.015}_{-0.013}$
Zhang et al. (2013)	K-dwarfs	0.0065 ± 0.0023
Bienaymé et al. (2014)	Red clump	0.0143 ± 0.0011
Xia et al. (2016)	K and G dwarfs	0.018 ± 0.0054
Sivertsson et al. (2018)	G-dwarfs	0.012 ± 0.002
Buch et al. (2018)	A-dwarfs	0.023 ± 0.010
	F-dwarfs	$0.047^{+0.006}_{-0.007}$
	Early G-dwarfs(G0-G3)	$0.021^{+0.014}_{-0.011}$

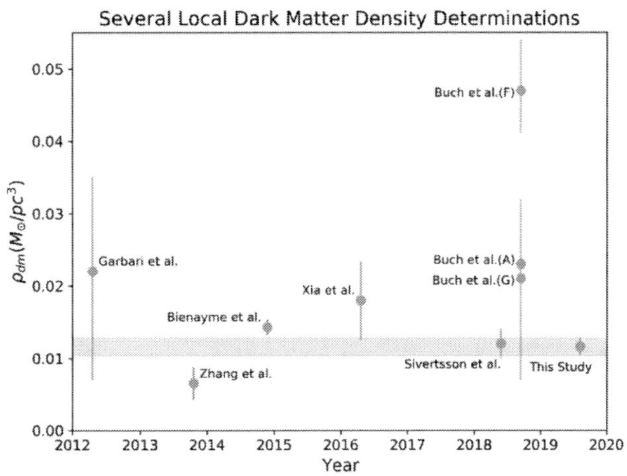

Figure 2. Comparison of our result with other determinations of ρ_{dm} conducted by some researchers. The error bar in our result takes into account only statistical uncertainties

1. **Different data used in determination**

Garbari et al. (2012)[4] used data from the catalogs in Kuijken & Gilmore (1989b), Zhang et al. (2013)[5] from SDSS-SEGUE observations, Bienaymé from RAVE observations, Xia et al. (2016)[7] from LAMOST observations, Buch et al. (2018)[9] used Gaia DR2, and Sivertsson et al. (2018)[8] from SDSS-SEGUE observations.

Different missions usually have different precisions and produce data with different errors. As the likelihood function uses the error values to calculate the posterior distribution, and the error is propagating, then differences in the data can cause differences in the posterior values obtained.

The selection function of each observation can also be different. Selection function can be related to the sensitivity of the instrument used. If the instrument sensitivity differs, it is possible that the objects observed are also different. Generally, instruments with higher sensitivity can observe more objects.

Missions with different locations can also generate different data because the location determines the area of the sky that can be observed. RAVE missions, for example, cannot observe objects in the northern hemisphere while the LAMOST missions cannot observe objects in the southern hemisphere. This causes objects that can be observed by both missions to be different.

2. **Different stellar population used as tracers.**

   Different stellar populations can be in different states with regard to dynamical equilibrium. Older stellar population is more likely closer to dynamical equilibrium state than the younger one. The difference in these conditions can cause differences in the results of searched parameters. The further the tracer is from a state of dynamical equilibrium, the more important the term $\partial f/\partial t$ in the collisionless Boltzmann equation is to be involved in calculation.

3. **Different volume and tracers' height considered**

   Garbari et al. (2012)[4] considered tracers around the South Galactic Pole with height of 200 pc - 2,000 pc. Zhang et al. (2013)[5] considered the height of 200 pc - 2,000 pc, whereas Bienaymé (2014)[6] used a cylindrical volume with a radius of 500 pc that orients to the South Galactic Pole and with a height of 200 pc to 2,000 pc. Xia et al. (2016)[7] took a cone-shaped volume with a diameter of 10° around the North Galactic Pole with an altitude of 200 pc - 1500 pc. Buch et al. (2018)[9] considered the same cylinder as this study, while Sivertsson et al. (2018)[8] took a radius 1 kpc cylinder with half-height of 3 kpc.

   The differences in the shape of the volume and the height considered cause the total mass in the volume being considered to also be different, and it can make the value of ρ_{dm} obtained to be different as well. On the other hand, all of these determinations assumed a constant value of ρ_{dm} in the considered volume. As ρ_{dm} is in fact not homogeneously distributed [1], the larger the space being reviewed, the less accurate the assumption of a constant ρ_{dm} is.

4. **Different applied assumptions**

   Garbari et al. (2012)[4] and Buch et al. (2018)[9] assumed that stars and gas are isothermal. By assuming that, the vertical velocity dispersion of stars and gas is assumed to be constant for all the z considered. In addition, Buch et al. (2018)[9] also added a dark disk component in the model. Meanwhile, the main equation in this study followed model for the vertical velocity dispersion of tracers as a function of height from the Galactic plane, as explained by Xia et al. (2016)[7]. The model is then fitted with velocity dispersion obtained from observational data by varying the value of the free parameters.

   Sivertsson et al. (2018)[8] made a model for the tilt term in Jeans equation. The model allows the determination of ρ_{dm} without neglecting the tilt term. Meanwhile, Bienaymé et al. (2014)[6] did not make the stellar disk scale-height, h_*, as a free parameter but rather made it as a constant value at 300 pc.

   The differences in these assumptions can result in different applied main equation, or different quantities used as free parameters, thus causing differences in the results of ρ_{dm} obtained.

5 Conclusion

We examined kinematics and positions of 95,543 K-dwarfs from Gaia DR2 to determine the local dark matter density. The value we acquire is 0.0116 ± 0.0012 $M_\odot/pc^3$ or $\rho_{dm} = 0.439 \pm$

0.046 GeV/cm^3, which still agrees with the results obtained by Garbari et al. (2012)[4], Xia et al. (2016)[7], Sivertsson et al. (2018)[8], and Buch et al. (2018)[9] in the range of errors. Our result is larger than those obtained through the decomposition of Milky Way's rotation curve by de Salas et al. (2009)[2] and Catena & Ullio (2010)[3] who get values of $\rho_{dm} \cong 0.3 - 0.4$ GeV/cm^3 and $\rho_{dm} = 0.385 \pm 0.027$ GeV/cm^3, respectively. This gives us a hint that the shape of our Galaxy's dark halo is oblate.

References

[1] Read, J. I., Journal of Physics G: Nuclear and Particle Physics **6**, 41 (2014)
[2] de Salas, P. F., et al., Journal of Cosmology and Astroparticle Physics **10**, id. 37 (2019)
[3] Catena, R, & Ullio, P., Journal of Cosmology and Astroparticle Physics **8**, id. 4 (2010)
[4] Garbari, S., et al., Monthly Notices of the Royal Astronomical Society **425**, 1445-1458 (2012)
[5] Zhang, L., et al., ApJ **772**, 108-121 (2013)
[6] Bienaymé, O., et al., Astronomy & Astrophysics **A92**, 571 (2014)
[7] Xia, Q., et al., Monthly Notices of the Royal Astronomical Society **458**, 3839-3850 (2016)
[8] Sivertsson, S., et al., Monthly Notices of the Royal Astronomical Society **478**, 1677-1693 (2018)
[9] Buch, J., et al., Journal of Cosmology and Astroparticle Physics **4**, id. 26 (2019)
[10] Gaia Collaboration, et al., Astronomy & Astrophysics **A1**, 595 (2016)
[11] Gaia Collaboration, et al., Astronomy & Astrophysics **A1**, 616 (2018)
[12] Binney, J., & Tremaine, S., *Galactic Dynamics*, (Princeton University Press, Princeton, 2008)
[13] Bovy, J., Monthly Notices of the Royal Astronomical Society **468**, 63-67 (2017)
[14] McKee, C. F., et al., ApJ **814**, 13-37 (2015)
[15] Poleski, R., arXiv:1306.2945v3 (2013)
[16] Pryor, C., & Meylan, G., ASP Conference Series **50**, 1993.
[17] Spark, L. S., & Gallagher, J. S., *Galaxies in the Universe an Introduction, Second Edition*, (Cambridge University Press, New York, 2007)
[18] Büdenbender, A., et al., Monthly Notices of the Royal Astronomical Society **452**, 956-968 (2015)
[19] Evans, D. W., et al., Astronomy & Astrophysics **A1**, 616 (2018)
[20] Silverwood, H., et al., Monthly Notices of the Royal Astronomical Society **459**, 4191-4208 (2016)

On the origin of high Doppler velocity wings in the spectra of O-rich AGB stars

Do Thi Hoai[1], *Pham Tuyet* Nhung, *Pham* Tuan-Anh, *Pierre* Darriulat, *Pham Ngoc* Diep, *Nguyen Thi* Phuong and *Tran Thi* Thai

Department of Astrophysics, Vietnam National Space Center, Vietnam Academy of Sciences and Technologies (VNSC/VAST), 18 Hoang Quoc Viet, Nghia Do, Cau Giay, Hanoi, Vietnam

Abstract. Millimetre ALMA observations of the nascent winds of several Oxygen rich AGB stars have revealed the high Doppler velocity wings in their spectra. However, the physics underlying their production is unclear. In this paper, we illustrate the argument with four examples of oxygen-rich AGB stars: EP Aqr, R Dor, L2 Pup and Mira Ceti.

1. Introduction

At the end of their life, when they have burned enough hydrogen into helium to produce a helium core in their centre, stars similar to our Sun blow up by some two orders of magnitude to become Red Giants. They then enter the Asymptotic Giant Branch (AGB) where they start losing mass in the form of a wind of gas and dust. At the same time nuclear reactions take place in the core and new elements are dredged up into the external layers of the star.

The expulsion of material in the interstellar medium is believed to proceed as follows: close to the stellar surface dust grains condense as soon as the temperature is low enough ($<10^3$ K or so). They get radial momentum by absorbing UV radiation from the star, which they transfer to gas molecules by collisions, enough to escape the star gravity. Dust formation depends very much on the C/O ratio in the gas atmosphere; most C and O atoms bind to produce CO molecules and whatever is left in excess defines whether one deals with an "oxygen star" or a "carbon star". Carbon-based grains are more effective than oxygen-based grains at absorbing the stellar radiation.

Dust is observed from black body radiation of grains at visible and infrared wavelengths either from space or from ground using adaptative optics and/or interferometry. Gas is best observed from the millimetre emission of CO molecules that are abundant and easily excited. Observations at millimetre wavelength provide information of the intensity projected on the sky plane and the Doppler velocity. Studies of the morpho-kinematics of O-rich AGB stars show that the gas wind reaches typically a terminal velocity of ~ 10 km s^{-1} at a distance of few hundreds au. However, faint higher velocity wings, in particular in SiO line profiles, have been observed on many of them.

1 Corresponding author: dthoai@vnsc.org.vn

The occasional presence of high velocity wings in single dish millimetre observations of the SiO line emission has been noticed for some time and has been analysed by Winters et al. [1] and more recently by De Vicente et al. [2]. They quote velocities as large as twice the terminal velocity and suggest that the emission occurs close to the star, where SiO grains have not yet fully formed, and is somehow related to star pulsations.

Recently, the availability of high angular resolution ALMA (Atacama Large Millimeter/sub millimeter Array) observations of several oxygen rich AGB stars like EP Aqr, R Dor, L_2 Pup and Mira Ceti have shed a new light on this issue [3-6]. However, the origin of these high velocities is still unclear. While EP Aqr seems to favour a scenario in terms of bipolar streams, L_2 Pup is naturally described as very high rotation velocities close to the star, while R Dor and Mira Ceti require different interpretations. In this paper, we summarise the state of the art of such studies.

2. EP Aqr

EP Aqr is an oxygen-rich star on the early phase of the AGB with no technetium in its spectrum [7] at a distance of 114±8 pc from the Sun [8-9]. It has a mass between 1 and 2 $M_{\odot}$, a mass loss rate of ~1.6 10^{-7} $M_{\odot}$ yr^{-1} [6] and an effective temperature of ~ 3200 K [10].

The line emission of different molecules has been observed, among which CO, SiO and SO_2 which probe preferentially different distances from the star. While CO molecules are formed early and survive up to long distances, being only affected by UV dissociation from the interstellar medium [11-12], SiO molecules tend to aggregate in grains, causing their progressive disappearance from the gas [13-14] at distances at 100 au scale in addition to their early dissociation from stellar UV radiation [15]. While both CO and SiO emissions are excited by collisions with other molecules, in majority hydrogen, SO_2 emission is mostly excited (but also dissociated) by absorbing the UV radiation of the star, confining the emission to its close neighbourhood [16-17]. Figure 1 shows the dependence on projected distance to the star, *R*, of the intensity multiplied by *R*. The beam sizes of these observations are 0.33×0.30 arcsec^2, 0.29×0.25 arcsec^2 and 0.18×0.17 arcsec^2 for $^{12}CO(2\text{-}1)$, SiO(5-4) and $SO_2(16_{6,10}\text{-}17_{5,13})$ respectively.

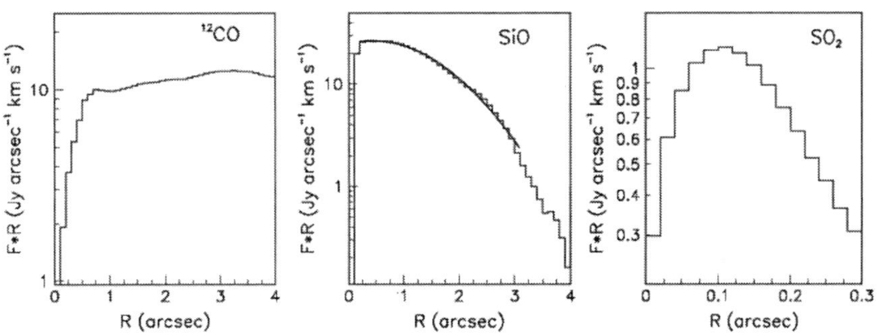

Figure 1: EP Aqr: Dependence on *R* of the intensity multiplied by *R* for CO, SiO and SO_2 (from left to right).

The CO emission at distances exceeding ~250 au [6] displays approximate axi-symmetry about an axis making an angle φ ~10° with the line of sight and projecting on the sky plane some 20° west of north. The Doppler velocity spectra are made of two components: a narrow component associated with an equatorial outflow expanding at low velocity V_{eq}=(0.33 ± 0.03)/sinφ km s^{-1} and a broad component associated with a bipolar outflow expanding at a velocity increasing with stellar latitude α up to 10 to 12 km s^{-1} near the poles. A possible rotation velocity at the equator is shown not to exceed a third of the expanding velocity.

SiO emission gives evidence for a high velocity (~20 km s^{-1}) wind close to the star. The P-V maps of Figures 2 show that the wind at short distances reaches a velocity well above the terminal velocity of ~12 km s^{-1} observed in CO emission at large distances from the star [6].

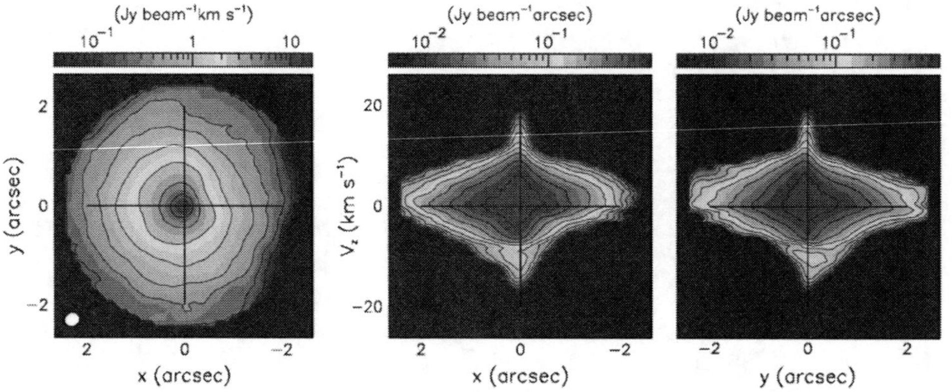

Figure 2: EP Aqr: Projections of the SiO data-cube on (x, y) (left), (x, V_z) (middle) and (y, V_z) (right).

SO_2 emission does not show any evidence for such high velocity components. Instead, the map of the mean Doppler velocity of SO_2 (Figure 3) suggests the presence of rotation. Distributions of the Doppler velocity as a function of position angle ψ measured counter-clockwise from the y axis (Figure 3 right) is well described by a sine wave; its phase shift nearly cancels, meaning that the axis of rotation projects on the sky plane along the y axis, implying the absence of a significant anisotropy, polar or equatorial, of the expansion velocity [18].

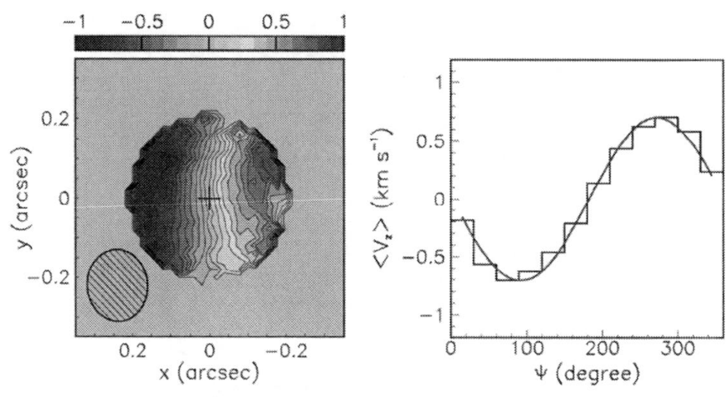

Figure 3: SO_2 emission of EP Aqr: sky map of the mean Doppler velocity (km s^{-1}, $R < 0.2$ arcsec) (left) and dependence of the Doppler velocity on position angle ψ for $0.05 < R < 0.15$ arcsec (right).

The absence of high velocities in SO_2 emission, if not the result of insufficient sensitivity, implies that the high velocity wings do not occur very close to the star. It also disfavours an interpretation in terms of a spherical shell ejected by star pulsation at short distances from the star, as described in Winters et al. [1] and McDonald & Zijlstra [19]. It favours instead an interpretation in terms of polar gas streams which is also consistent with the observations at large distances of CO emission which display clear bipolarity with a factor of 6 between polar and equatorial winds [6].

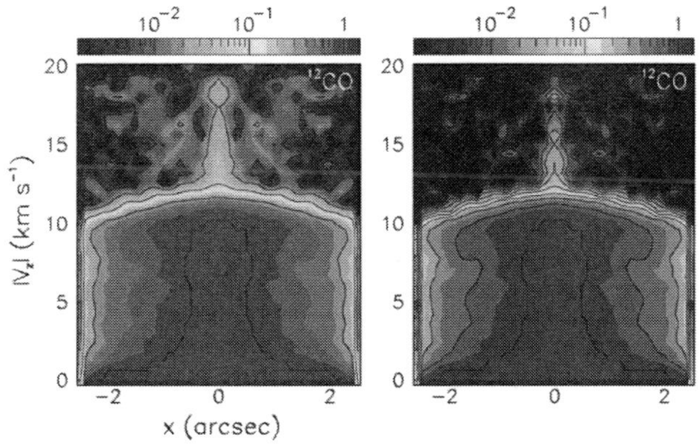

Figure 4: P–V maps of the intensity (Jy arcsec^{-1}) for centrally symmetrized CO observations for a 1.5σ noise cut (left) and for a 2σ cut (right).

The high velocity streams are present in CO emission but weaker than in SiO emission (Figure 4), suggesting that they slow down and/or diverge at large distances from the star. These streams are very narrow (~10°), launched from less than 25 au from the star on the line of sight direction and build up between ~20 au and ~100 au to a velocity of ~ 20 km s^{-1} [3]. The mechanism governing the launching of the observed streams remains unknown. However, the coincidence of the stream axis with the symmetry axis of the circumstellar envelope and the SO_2 rotation axis strengthens the interpretation of high Doppler velocity wings in terms of gas streams.

3. L_2 Pup

L_2 Puppis is one of the nearest AGB stars at a distance of 64 ± 4 pc [20]. It is a semi-regular pulsating variable (P ~140 days, [21-22]). An asymmetric resolved environment around L_2 Pup was first identified by Ireland et al. [23] using aperture masking in the optical and near-infrared. An edge-on disk with an inclination angle of 82° has been observed by Kervella et al. [24] in 2014 using VLT/NACO adaptive optics between 1.0 and 4.0 μm. It was later confirmed by Lykou et al. [25-26], Ohnaka et al. [27] and Kervella et al. [28] in 2015. Kervella et al. [28] identified the inner rim of the disk through its polarimetric signature at a radius of 6 au from the AGB star. They revealed a close companion at a projected separation of 2 au from the main star in R and V band VLT observations, detected bipolar "hourglass" cones in L_2 Pup's envelope and discovered streamers in the bipolar cones and two thin, tightly collimated plumes.

The companion is also detected from the continuum and molecular line emissions of ALMA band 7 observations [28]. It has a mass of 12 ± 16 M_{Jup} and is most likely a planet or a low-mass brown dwarf with an orbital period of about five years.

In contrast with EP Aqr, L_2 Pup displays strong rotation around an axis in the plane of the sky ([4], Figure 5 and 6). It is approximately Keplerian within the central cavity of the dust disk. This allows for estimating the mass of the main star as ~0.66 solar masses. The nature of the rotation also implies very high rotation velocities close to the star, ~20 km s^{-1} at 1.3 au, that appear in projection close to the line of sight. It is therefore natural to interpret the origin of the high Doppler velocities observed close to the star in terms of rotation.

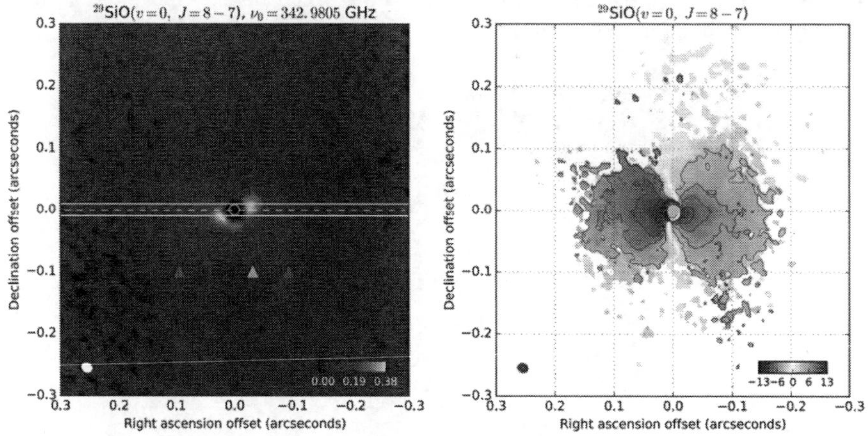

Figure 5: Left panel: map of the emission from L2 Pup in the ^{29}SiO(v=0, J=8−7) line. The solid white lines present the slit of 20 mas used to compute the position-velocity diagrams. Right panel: Doppler velocity map (the colour scale is in km s^{-1}).

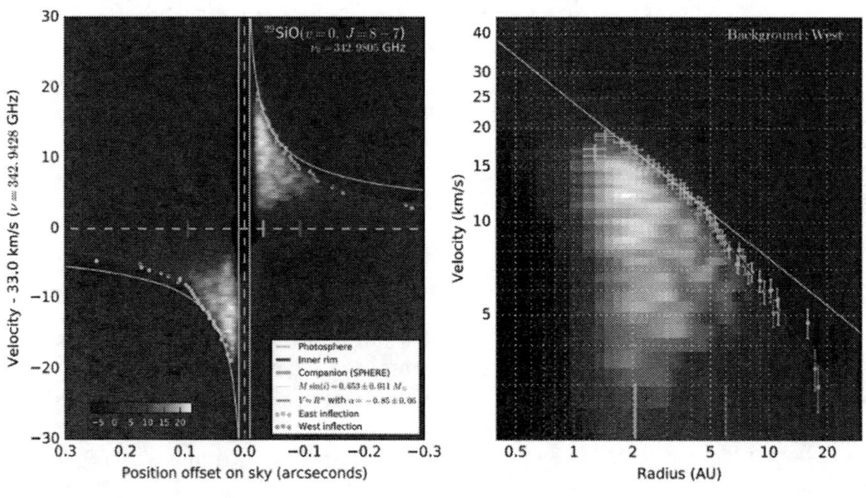

Figure 6: L2 Pup: Position-velocity diagrams plotted inside the slit shown in Figure 5. The yellow curve is a Keplerian fit to the inner cavity, and the red curve is a power law fit to the non-Keplerian radius domain.

4. R Dor

R Dor is an oxygen rich AGB star very similar to EP Aqr in terms of spectral classes, masses, mass loss rates and temperatures. It also has no technetium in its spectrum [30]. It is the closest AGB star to the Sun, at distance of ~59 pc [29]. It has a dual pulsation period of 175 and 332 days [30].

The circumstellar envelope of R Dor has been probed by ALMA with a resolution of ~4 au up to some 60 au from the star [31-33] and the dust has been observed at the Very Large Telescope (VLT) with a resolution of 1.2 au [34]. In addition, below ~15 au, the analyses of Danilovich et al. [16], De Beck & Olofsson [35] and Van de Sande et al. [14] have contributed a considerable amount of detailed information of relevance to the physicochemistry and dynamics of both dust and gas. At larger distances from the star, an analysis of ALMA observations of SO(J_K=6_5−5_4) emission [36], probing distances between 20 and 100 au, gives evidence for the wind to host a radial outflow covering large solid angles and displaying strong inhomogeneity both in direction and radially: the former takes the form of multiple cores and the latter displays a radial dependence suggesting an episode of enhanced mass loss having occurred a century or so ago.

Decin et al. [31] were first to discuss the presence of high velocity wings, far above the canonical terminal wind velocities, not only in the SiO but also in the CO line emissions from the vibrational ground state, with too low an excitation energy to be caused by pulsations [37]. They argue that neither star pulsations nor thermal motions of the gas can significantly contribute to the generation of such high velocities.

Hoai et al. [5] made a more detailed and dedicated study of this issue; they use line emissions detected between ~15 and ~50 au from the star, including five different molecular excitations: CO(3-2), SiO(8-7), SO(6_5-5_4), SO_2($13_{4,10}$-$13_{3,11}$) and HCN(4-3) with an angular resolution of ~0.16 arcsec.

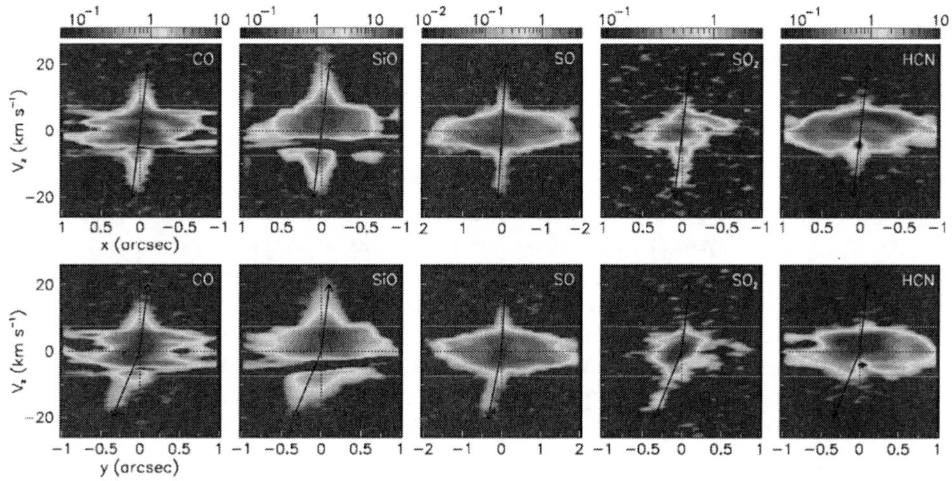

Figure 7: R Dor: P-V maps in the V_z vs x and vs y planes. The colour scale is in units of Jy arcsec^{-1}. Yellow lines show the cuts applied in the definition of the large Doppler velocity components.

Projections of the data-cubes on the (x, V_z) and (y, V_z) planes (P-V maps) are shown in Figure 7. They are the integration over y, pointing 35° west of north, and x, pointing 35° north of east, respectively. In all cases, the larger values of $|V_z|$ are confined near $x = 0$, very much as was observed in EP Aqr. They are particularly visible in the CO, SiO and SO maps, both in the blue-shifted and red-shifted hemispheres, but more clearly in the former than in the latter.

If these high velocity winds are ejected in the form of gas streams and reach velocity of 20 km s^{-1} at a same distance in space like in the case of EP Aqr, the red-shifted and blue-shifted streams should be both confined near the line of sight and nearly back-to-back. However, in variation to EP Aqr, the rotation axis observed by Homan et al. [32] makes an angle of only 20 ± 20° with the plane of the sky, far from the line of sight. This may suggest that the observed high Doppler velocity wings are artefacts of improper data reduction and/or continuum subtraction; however, the Doppler velocity distribution of continuum-unsubstracted data (Figure 8) shows that such is not the case: the high velocity wings are still present in continuum-unsubstracted data. Blue-shifted and red-shifted high velocity components display amazing similarity in spite of some expected absorption.

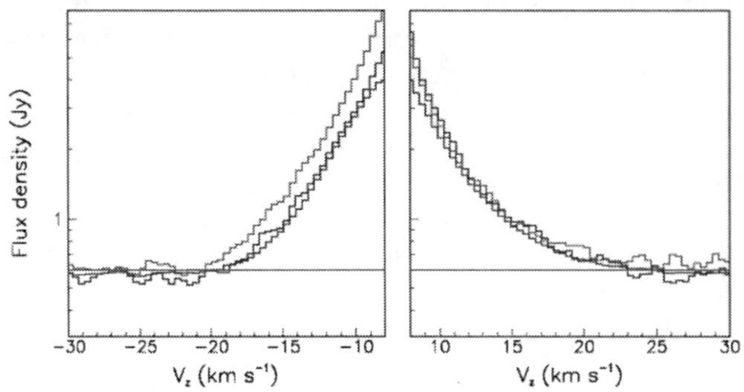

Figure 8: Doppler velocity distributions obtained for $|V_z| > 8$ km s^{-1} using two data sets of SiO line emission having significantly different uv coverage. The data set having the larger maximal recoverable scale is shown for $R < 1.5$ arcsec (red) and for $R < 0.3$ arcsec (blue). The data set having the smaller maximal recoverable scale is shown for $R < 1.5$ arcsec (black).

Interpretation in terms of rotation is difficult to justify in this case. Homan et al. [32] quote a rotation velocity of only 12 km s^{-1} at a distance of 6 au from the star. Keplerian rotation would imply that the maximal observed velocity of 20 km s^{-1} would be reached at only ~2.2 au from the star centre, very close to the stellar surface. The absence of detection in the ^{28}SiO(v=1, J=8-7) data analyzed by Homan et al. [32] would then have to be blamed on insufficient sensitivity if a Keplerian rotation scenario were retained. Vlemmings et al. [33] quote a solid body rotation of 1.0±0.1 km s^{-1} at the stellar surface (~1.9 au) meaning that a rotation velocity of 20 km s^{-1} would be reached at a distance of ~38 au from the star, nearly twice as far as the canonical limit that has been set. Unfortunately, the authors do not comment on the high Doppler velocity wings in general. They only illustrate the blue-shifted stream, which they claim to be one-sided, with a P-V diagram of ^{29}SiO(v=0, J=5−4) data that is consistent with the gas stream interpretation. However, both rotation and pulsation predict red-blue symmetry in contrast with what is observed.

5. Mira Ceti

Mira Ceti is one of the most studied AGB stars, close to the Sun ~100±10 pc [9, 38]. It is remarkable for the large amplitude of its variability, ~8 mag, with a periodicity of ~333 days [39]. The direct evidence for a companion was first obtained in 1995 by the Hubble Space Telescope in visible light [40]. Its orbit is poorly constrained by observations [41]: if circular its projected eccentricity corresponds to an angle of ~22° with the line of sight, the semi-major axis being ~ 0.8 arcsec. The orbital period is estimated to be at least 500 years. While the companion has been resolved at many other wavelengths, its nature is still a matter of debate, some claiming that it is a White Dwarf [42-44] and others claiming that it is a Main Sequence star [45].

Recent studies using high resolution observations of ALMA at radio wavelengths have shown the complex morpho-kinematics of the circumstellar envelope of Mira AB [6,46-47].

Close to the stars ($R < 0.25$ arcsec), large values of $|V_z|$ reaching 16 km s^{-1} for both CO and SiO have been revealed (Figure 9). They are more clearly marked when centred on the main star, Mira A, than on its companion, Mira B. This observation raises the question of the origin of such large Doppler velocity wings.

Since there is no evidence for the line of sight to play a particular role in the dynamics of the Mira AB pair, it would be unreasonable to retain an interpretation in terms of gas streams. Moreover no significant rotation is detected near Mira A, leaving stellar pulsation as a candidate interpretation.

However, CO and SiO observations have been made at very different phases of the Mira pulsation cycle [48-49], CO near maximum and SiO near minimum. The radial velocity of the stellar pulsation is expected to span some 20 to 30 km s^{-1} between maximum and minimum [37]: the velocities measured in CO and SiO emissions should be different, in contradiction with observations.

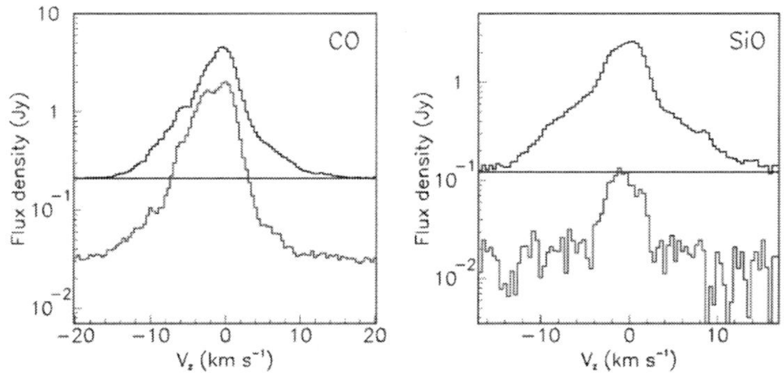

Figure 9: Mira Ceti: Doppler velocity distributions of CO (left) and SiO (right) line emission within a circle of 0.25 arcsec projected radius centred on the main star Mira A (black, upper) or its companion Mira B (red, lower).

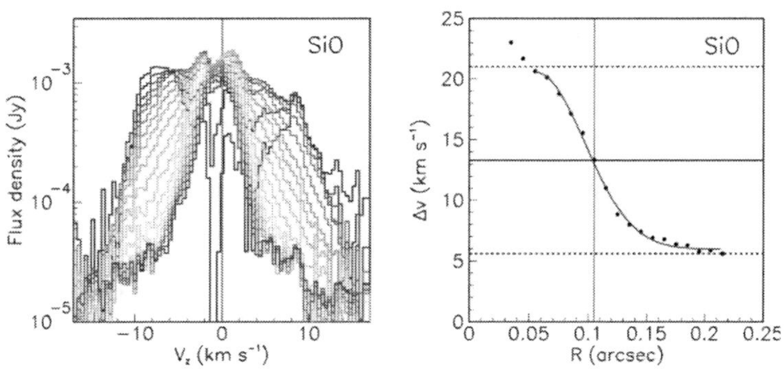

Figure 10: Mira Ceti: Left: Doppler velocity spectra of SiO averaged in 0.03 arcsec wide rings centred on Mira A with mean radii increasing from 0.025 arcsec (blue) to 0.215 arcsec (red) in steps of 0.01 arcsec. Right: dependence on *R* of the full-with at 1/5 maximum of the spectra displayed in the centre-right panel.

The spectra of SiO in different rings centred on the star (Figure 10) are broader close to the star and narrower at larger distances with a rather sharp transition at ~0.1 arcsec. It suggests that these large velocities are reached at typical distances of 5 to 10 au from Mira A and are caused by a large line width rather than by a large wind velocity. As thermal broadening cannot produce such large line widths, turbulence must be invoked. The picture would then be of a shell surrounding the zone where stellar pulsation dominates and extending up to ~10 au, hosting a very turbulent regime. Indeed, such a possibility was mentioned earlier by Kamínski et al. [50] in relation with the possible presence of very broad AlO line profiles. According to the standard picture of acceleration by transfer of momentum from dust grains to molecules of gas([51], this is a region that hosts important shocks. At 10 au, the escape velocity from one (two) solar masses, values commonly accepted for Mira A, is ~13 (19) km s^{-1} and the density is such that the velocity acquired by a gas molecule after its interaction with a dust grain is immediately thermalized. Such important turbulence has been observed in a different context by Falgarone & Phillips [52] and discussed by Gillet et al. [53] in the case of pulsating stars.

6. Summary

The presence of large Doppler velocities in molecular line emission spectra of oxygen-rich AGB stars has been observed for quite some time but is still not thoroughly understood. We have discussed here the cases of four stars close enough to the Sun to have lent to high angular resolution measurements. However, these observations suggest interpretations that differ from star to star, making one be suspicious of their ad hoc nature. The interpretation as gas streams favoured in the case of EP Aqr seems difficult to extend to the cases of the other stars, for which the line of sight plays no special role. The interpretation in terms of rotation favoured in the case of L_2 Pup has been shown not to apply to the other stars. Simple interpretations in terms of stellar pulsations have been shown to be untenable quantitatively. To the extent that such large Doppler velocity wings share many properties, it seems desirable to find a common interpretation that could apply to all cases. The possible importance of turbulence in a layer surrounding the pulsation region revealed by the study of Mira Ceti needs to be better understood. A global analysis of all available observations of relevance, together with additional high angular resolution observations are required to better sort out the arguments of relevance and deepen our understanding of the mechanisms at stake.

We are deeply indebted to the ALMA partnership, whose open access policy means invaluable support and encouragement for Vietnamese astrophysics. Financial support from the World Laboratory, VNSC and NAFOSTED is gratefully acknowledged. This research is funded by the Vietnam National Foundation for Science and Technology Development (NAFOSTED) under grant number 103.99-2018.325.

Reference

1. Winters J. M., Le Bertre T., Jeong K. S., Nyman L.-Å., Epchtein N., A&A, **409**, 715 (2003)
2. De Vicente P. et al., A&A, **589**, A74 (2016)
3. Tuan-Anh et al., MNRAS, **487**, 622 (2019)
4. Kervella P., Homan W. , Richards A. M. S. et al., A&A, **596**, A92 (2016)
5. Hoai et al., Comm. Phys. Vietnam, **20**, 85, 3 (2020), aXiv: 1906.08535
6. Hoai D. T. et al., MNRAS, **484**, 1865 (2019)
7. Lebzelter, T., and Hron, J., A&A, **351**, 533 (1999)
8. Gaia Collaboration, A&A, **616**, A1 (2018)
9. van Leeuwen F., A&A, **474**, 653 (2007)
10. Dumm T., & Schild H., New Astr., **3**, 137 (1998)
11. Groenewegen M. A. T., A&A, **606**, A67 (2017)
12. Mamon G. A., Glassgold A. E., Huggins P. J., ApJ, **328**, 797 (1988)
13. Gonzalez Delgado D., Olofsson H., Kerschbaum F. et al., A&A, **411**, 123 (2003)
14. Van de Sande M., Decin L., Lombaert R. et al., A&A, **609**, A63 (2018)
15. Willacy K., Millar T. J., A&A, **324**, 237 (1997)
16. Danilovich T., De Beck E., Black J. H., Olofsson H., Justtanont K., A&A, **588**, A119 (2016)
17. Yamamura I., de Jong T., Onaka T., Cami J., Waters L. B. F. M., A&A, **341**, L9 (1999)
18. Diep P. N., Phuong N. T., Hoai D. T. et al., MNRAS, **461**, 4276 (2016)
19. McDonald I., Zijlstra A. A., ApJ, **823**, L38 (2016)
20. van Leeuwen F., A&A, **474**, 653 (2007)
21. Bedding, T. R., Kiss, L. L., Kjeldsen, H., et al., MNRAS, **361**, 1375 (2005)

22. Kholopov, P. N., Samus, N. N., Kazarovets, E. V., & Perova, N. B., Information Bulletin on Variable Stars, **2681**, 1 (1985)
23. Ireland, M. J., Tuthill, P. G., Bedding, et al., MNRAS, **350**, 365 (2004)
24. Kervella, P., Montargès, M., Ridgway, S. T. et al., A&A, **564**, A88 (2014)
25. Lykou, F., Klotz, D., Paladini, C. et al., A&A, **576**, A46 (2015)
26. Lykou, F., Klotz, D., Paladini, C. et al., A&A, **581**, C2 (2015)
27. Ohnaka, K., Schertl, D., Hofmann, K.-H. & Weigelt, G., A&A, **581**, A127 (2015)
28. Kervella, P., Montargès, M., Lagadec, E., et al., A&A, **578**, A77 (2015)
29. Knapp G.R., Pourbaix D., Platais, I., and Jorissen, A., A&A, **403**, 993 (2003)
30. Bedding T. R., Zijlstra A.A., Jones A., & Foster G., MNRAS, **301**, 1073 (1998)
31. Decin L., Richards A.M.S., Danilovich T., et al., A&A, **615**, A28 (2018)
32. Homan W., Danilovich T., Decin L., et al., A&A, **614**, A113 (2018b)
33. Vlemmings W.H.T., Khouri T., De Beck E., et al., A&A, **613**, L4 (2018)
34. Khouri T., Maercker M., Waters L.B.F.M., et al., A&A, **591**, A70 (2016)
35. De Beck E. & Olofsson H., A&A, **615**, A8 (2018)
36. Nhung P.T., Hoai D.T., Tuan-Anh P., et al., MNRAS, **490**, 3329 (2019)
37. Nowotny, W., Höfner, S. and Aringer, B., A&A, **514**, A35 (2010)
38. Haniff, C. A., Scholz, M., & Tuthill, P. G., MNRAS, **276**, 640 (1995)
39. Templeton M. R. & Karovska M., 2009, ApJ, **691**, 1470 (2009)
40. Karovska, M., Hack, W., Raymond, J., & Guinan, E., ApJ, **482**, L175 (1997)
41. Prieur, J. L., Aristidi, E., Lopez, B. et al., ApJS, **139**, 249 (2002)
42. Reimers, D. & Cassatella, A., ApJ, **297**, 275 (1985)
43. Wood, B. E. & Karovska, M., ApJ, **649**, 410 (2006)
44. Sokoloski, J. L. & Bildsten, L., ApJ, **723**, 1188 (2010)
45. Ireland, M.J., Monnier, J.D., Tuthill, P.G. et al., ApJ, **662**, 651 (2007)
46. Ramstedt, S., Mohamed, S., Vlemmings, W.H.T., et al., A&A, **570**, L14 (2014)
47. Nhung, P.T., Hoai, D.T. Diep, P.N. et al., MNRAS, **460**, 673 (2016)
48. Planesas, P., Alcolea, J. & Bachiller, R., A&A, **586**, A69 (2016)
49. Wong, K.T., Kamínski, T., Menten, K.M. & Wyrowski, F., A&A, **590**, A127 (2016)
50. Kamínski, T., Wong, K.T., Schmidt, M.R. et al., A&A, **592**, A42 (2016)
51. Höfner, S. & Olofsson, H., A&ARev., **26**, 1 (2018)
52. Falgarone, E. & Phillips, T.G., ApJ, **359**, 344 (1990)

Pulsations in Algols

David Mkrtichian[1,*], *Khemsinan* Gunsriwiwat[2], *Chris* Engelbrecht[3], *Napaporn* A-thano[4], *Victor* Nazarenko[5]

[1]National Astronomical Research Institute of Thailand, 260 Moo 4, T. Donkaew, A. Maerim, Chiangmai, 50180, Thailand

2 Department of Physics and Materials Science, Faculty of Science, Chiang Mai University, Thailand

3. Department of Physics, University of Johannesburg, PO Box 524, Auckland Park, 2006, South Africa

4. Institute of Astronomy, National Tsing Hua University, Hsinchu 30013, Taiwan

5. Astronomical Observatory, Odessa National University, Shevchenko Park, 65014, Odessa, Ukraine

Abstract. We present a brief review of the recent results of photometric and spectroscopic surveys and 3-D hydrodynamic simulations of the mass-transfer for the class of oscillating mass-accreting components of Algols (oEA stars). These ground-based spectroscopic and space-based photometric surveys are aimed at detecting and studying the spectra of non-radial oscillations, to get precise orbits and parameters of the binaries, and studying the spectroscopic and the hydrodynamic nature of the mass-transfer. We discovered 47 new oEA stars in TESS data and carried out follow up spectroscopic observations of known and new oEA stars with the SALT telescope. We found that the co-existence of excitation of high-degree non-radial modes with a wide spectrum of low-degree modes is a common phenomenon in oEA stars.

1 THE CLASS OF oEA STARS

The class of oEA pulsators was introduced by Mkrtichian et al. [1, 2]. These are components of semi-detached eclipsing Algol-type binary systems with mass transfer. The A or F-type mass-accreting primaries in these systems lie in the instability strip and show δ Scuti-like pulsations. This means that the evolution of this group of pulsators inside the δ Sct instability strip is different from classical δ Sct-type stars in detached eclipsing binary systems. The unique property of oEA stars is the co-existence of global stellar oscillations with accretion. Non-stationary accretion keeps them in thermal imbalance and drives their evolution along the Main Sequence over evolutionary tracks that are different from those of classical δ Scuti stars. Mkrtichian et al. [2] suggested that the episodes of non-stationary and rapid mass transfer and accretion on the surface of the pulsating gainer can potentially change the oscillation properties and modal spectra. Non-stationary mass transfer can be forced by cyclic magnetic activity of the cool Roche lobe filling component. According to this hypothesis, the pulsational properties and modal spectrum of oEA stars can be cyclically variable over time. A decade-long monitoring of the bright active oEA star RZ Cas confirmed this prediction [3]. Therefore, Algol-type systems with oEA components have a high potential when studying the physics of mass-transfer using

* Corresponding author: davidmkrt@gmail.com

asteroseismology, and there is a need to discover the diversity of such systems with different orbital and component parameters. By 2018, more than 70 oEA-type systems had been found by Mkrtichian, et al. [3], mostly by photometry.

2. 3-D HYDRODYNAMIC SIMULATIONS OF THE MASS-TRANSFER IN ALGOLS

In Algol systems, the flowing matter that escapes the Roche lobe-filling K-giant via the Lagrange L1 point is directly hitting and accreting onto the surface of the oscillating gainer in short-period systems or accreting onto the surface from the accretion discs formed in a long period system. The physics of the interaction of the gas stream with the atmosphere of the pulsating star is quite complex. To find the physical model that explains the spectroscopic peculiarities in the observed spectra of the component we need a 3-D model of the structure of the gas flow, mass accretion, and formation of the accretion disk in the large space around the studied binary system. We undertook a series of 3D-hydrodynamic simulations for different oEA systems observed during THASSOS and SALT surveys. The calculations were made using the version 3-D code developed by Nazarenko and Glazunova [4], which is based on the Large Particle Method developed earlier by Belotserkovskiĭ and Davydov [5]. Figure 1 shows some results of 3-D simulations obtained for the RZ Cas, TW Dra and AS Eri systems. For each system, we modelled different mass-transfer rates in order to find the structural differences in circumbinary gas flow and shock waves. We found almost conservative mass-transfer, with a mass-loss rate of less than one percent of the total mass transfer rate. An increase in the mass transfer rate changes the density of the gas stream and the structure of the circumbinary belt-like accretion annulus. These structural differences in the gas stream and circumbinary structures can explain the observed photometric and spectroscopic effects, namely orbital variations in pulsation amplitudes, the timely variable asymmetric Rossiter effect, and the variable intensity of He I lines reported in the RZ Cas and AS Eri systems by Nazarenko and Glazunova [4] and Mkrtichian et al. [11].

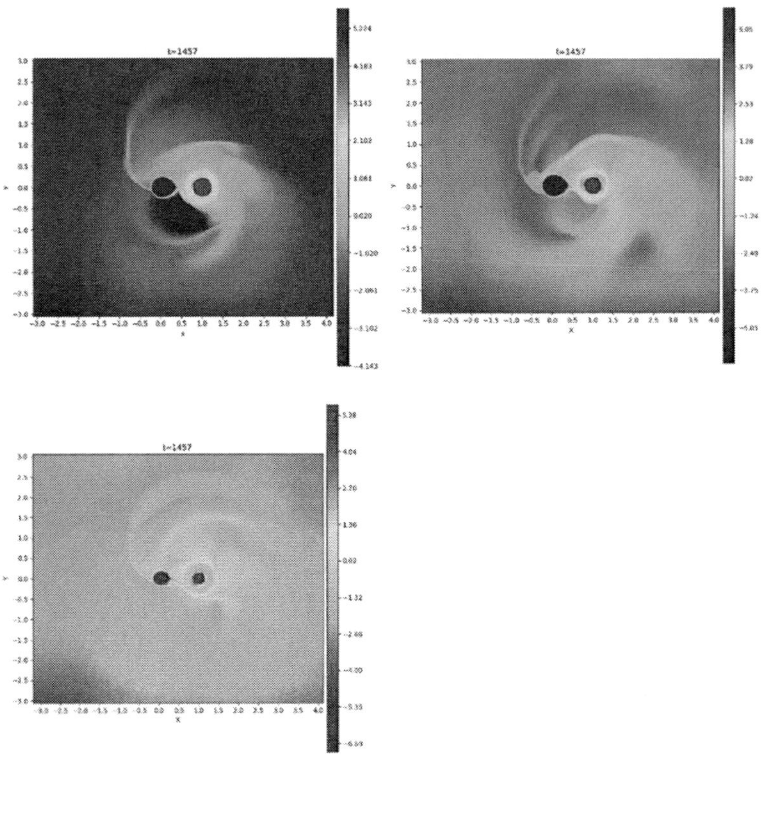

Fig. 1. *3-D hydrodynamic simulations of the mass-transfer and gas flow in the oEA systems RZ Cas (left), TW Dra (center) and AS Eri (right). This figure shows the particle density scale and the orbital plane (X-Y) cross section, based on Mkrtichian and Nazarenko.[6]).*

3 THE THASSOS SURVEY

Since the discovery of the class, >70 oEA-type systems have been found [3], mostly via ground-based photometry. In 2014–2019 at NARIT we carried out the photometric and the spectroscopic THAi Sky Survey of oEA Stars (THASSOS) in order to detect new oEA stars and to study their orbital, stellar, and pulsation parameters. For the photometric search for pulsations, we used the Thai Robotic Telescope Network of 70 cm telescopes located at different geographic longitudes (in Australia, China and the USA), and 60 cm Thai telescope PROMPT8 at CTIO (Chile). In total, 9 Algol-type systems were systematically monitored photometrically, and we discovered short-period pulsations in four of them, namely GQ TrA [7], VY Hya [8], UW Vir [9] and TT Vel [10]. For the THASSOS spectroscopic survey, we used the fiber-fed medium resolution (R=20,000) MRES echelle-spectrometer attached to the 2.4m Thai National Telescope at Mt. Doi Intanon (2450m). Our survey target list consisted of 15 known oEA systems, and for the majority of them we got complete orbital phase coverage. In particular, we got precise orbits for AS Eri, RX Hya and R CMa; the analysis of the spectra of the remaining target stars is ongoing.

4 THE TESS ALL-SKY SURVEY FOR oEA STARS

On 19 April 2018 NASA's Transiting Exoplanet Survey Satellite (TESS) was launched. During the 2018–2020 primary mission, TESS observed southern and northern hemisphere stars with its four wide-field cameras. We used the TESS 2-min cadence light curves of Algol stars for a massive all-sky survey aimed at detecting and studying new oEA stars. The open-access TESS light curve data were prepared by TESS Science Team, and we downloaded them from the Mikulski Archive for Space Telescopes (MAST). The out-of-eclipse light curves of all Algol-system with A-F primary components were consecutively analyzed. The pulsation spectra of newly discovered oEA stars have been analyzed in detail. The survey is ongoing, and detailed results of the discoveries of individual systems will be published in forthcoming papers. Here we briefly report preliminary results. To date, using the TESS data we have detected and classified 47 new oEA and 10 new δ Scuti stars in eclipsing binary systems. Figure 2 shows part of the light curve of the newly discovered oEA star V438 Vel. We investigated the pulsational spectra of previously known pulsators in AS Eri [11] and U Gru [12], and currently other discovered oEA stars are under analysis.

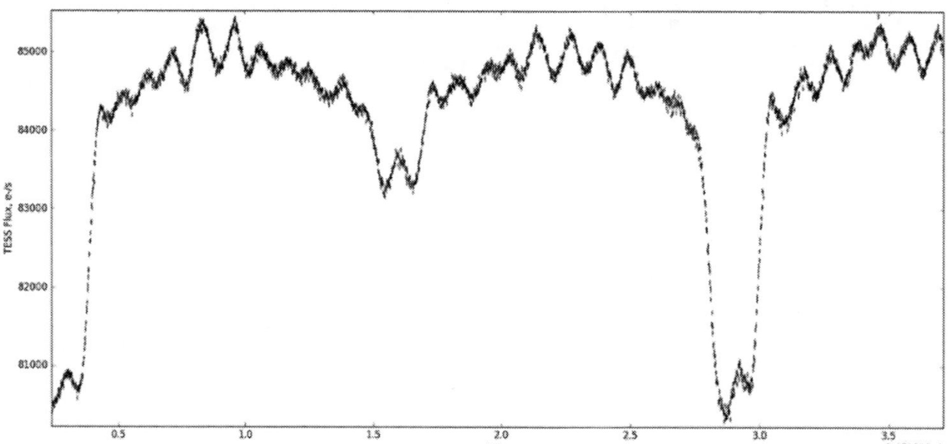

Fig. 2. The light curve of V438 Vel from the TESS space telescope. The high amplitude 0.132-day pulsations are clearly visible.

5 THE SALT SPECTROSCOPIC SURVEY OF SOUTHERN oEA STARS

The oEA stars are expected to be seen equator-on, which is the optimal inclination for the detection of sectoral (l = m) non-radial pulsations that have their greatest amplitudes around the stellar equator. Note that these high-degree (l,m>5) modes in photometric light variations have several orders of magnitude lower apparent amplitudes compared to low-degree mode (l,m<4) and are almost invisible in disk-integrated light. Indeed, the high-degree modes can be detected spectroscopically through variations of rotationally broadened line profiles. The detection of high-degree oscillation modes is important for asteroseismic analyses, as the small rotational period variations of pulsating mass-accreting components can be precisely measured using the Doppler effect [3]. These apparent variations have an effect on pulsation frequencies and are m-dependent. Thus, using variations in the frequencies of high-degree modes the accretion-driven timely variations in the rotation of the outer envelope can be precisely measured.

Our high-resolution spectroscopic survey of Southern oEA stars began in 2017 using the South African Large Telescope (SALT) at Sutherland Observatory, South Africa. We used the dual-beam fiber-fed high-resolution (R=40,000) echelle spectrograph HRS. The purpose of this on-going project is to extend the spectroscopic studies of non-radial pulsations to fainter equatorial and southern oEA stars; to find their accurate binary orbits and the parameters of the components; to detect and identify high-degree modes invisible in photometric studies; and to study the spectroscopic effect of the mass-transfer and accretion.

High-resolution spectroscopic time series of oEA stars have been reduced and analyzed using the Least Square Deconvolution (LSD) code of Glazunova et al. [16], allowing us to get a mean spectral line profile and increase the signal to noise ratio. The line profiles have been analyzed for the existence of non-radial pulsations. The survey data and the mode identification are still under investigation. Figure 3 shows the typical high-degree non-radial pulsation in line profile variations of oEA star VY Hya discovered during THASSOS survey.

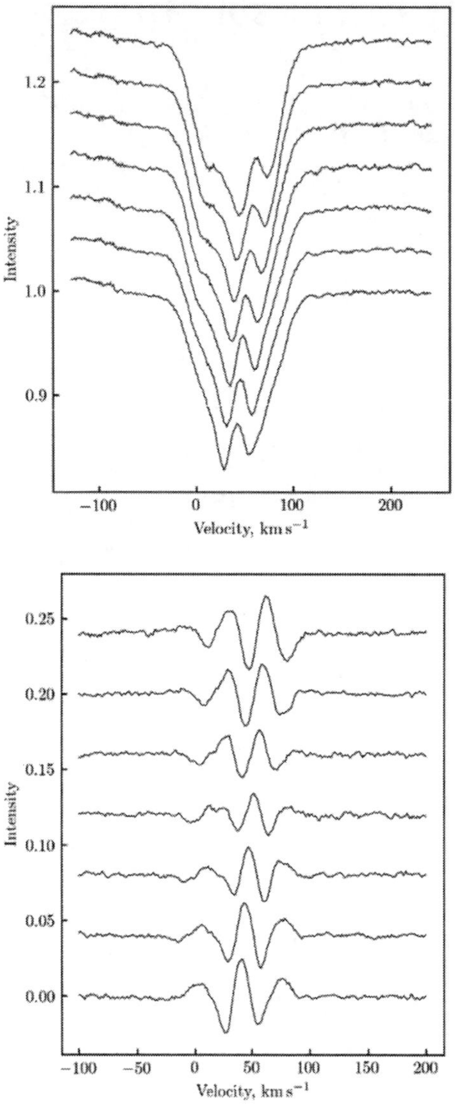

***Fig.** 3. The discovery of high-degree prograding non-radial mode pulsations in VY Hya. Left: Line-profile variations in LSD profiles. Right: Line-profile variations in the residual profiles.*

6 CONCLUSIONS

We carried out photometric and spectroscopic ground-based surveys and used TESS space telescope data for an all-sky search for new oEA stars. We discovered about 50 new oEA stars and did follow up spectroscopic observations of new oEA stars with the 2.4m Thai National and the SALT telescopes. We found that the excitation of high-degree non-radial pulsations is a common phenomenon and co-exists with a wide spectrum of low-degree modes in oEA stars. We carried out 3-D hydrodynamic simulations for several oEA systems in order to study the influence of the mass-transfer rate on the spatial structure of gas flows in the systems.

References

1. D. E. Mkrtichian, A. V. Kusakin, A. Y. Gamarova, & V. Nazarenko, PASPC, **259**, 96 (2002)
2. D. E. Mkrtichian, A. V. Kusakin, E. Rodriguez et al., A&A, **419**, 1015 (2004)
3. D. E. Mkrtichian, H. Lehmann, E. Rodriguez et al., A&A **475**, 4745 (2018)
4. V. V. Nazarenko, & L.V. Glazunova, ARep **57**, 294 (2013)
5. O. M. Belotserkovskiĭ, & Yu. M. Davydov, *Large Particles in Gas Dynamics,* Nauka, Moscow (1982).
6. D.E. Mkrtichian, & V. Nazarenko, n.d. (in preparation)
7. D. E. Mkrtichian, K. Gunsriwiwat, & S. Komonjinda, IBVS, **6182**, 1 (2016)
8. D. E. Mkrtichian, N. A-thano, S. Awipan, et al., IBVS, **6210**, 1 (2017)
9. D. E. Mkrtichian, K. Gunsriwiwat, S. Awipan, et al., IBVS, **6221**, 1 (2017)
10. D. E. Mkrtichian, K. Gunsriwiwat, D. Reichart, et al., IBVS, **6238**, 1 (2018)
11. D. E. Mkrtichian, K. Gunsriwiwat, C. Engelbrecht, et al., *Stars and Their Variability Observed from Space,* Eds.: C. Neiner, W. Weiss, D. Baade, C. Lovekin, & E. Griffin, in press.
12. D. Bowman, C. Johnston, A. Tkachenko, et al., ApJL, **883**, 26 (2019)
13. K. Gunsriwiwat, & D. E. Mkrtichian, IBVS, **6178**, 1 (2017)
14. D. E. Mkrtichian, C. Engelbrecht, P. Lampens et al., BSRSL, **88**, 256 (2019)
15. K. Gunsriwiwat, & D. E. Mkrtichian, IBVS, **6148**, 1 (2015)
16. L. V. Glazunova, A. V. Yushchenko, V. V. Tsymbal, et al., AJ, **136**, 1736 (2008)

 https://doi.org/10.1051/epjconf/202024005003

Hot Magnetic Stars in Exotic Multiple Systems

Eugene Semenko[1,2,*]

[1]National Astronomical Research Institute of Thailand (NARIT), 260 Moo 4, T. Donkaew, A. Maerim, Chiangmai, 50180 Thailand
[2]Special Astrophysical Observatory of the Russian Academy of Sciences, Nizhnii Arkhyz, Russia, 369167

Abstract. A strong and stable magnetic field covering the whole surface is a landmark of chemically peculiar or CP stars. To explain the origin of magnetic stars, several hypotheses were proposed, which depending on the mechanism and time of formation can be grouped into three main. The magnetic field may originate in the protostellar medium ("fossil field"), turbulent layers of a star ("dynamo" mechanism) or in the other environment. In either scenario, binary stars appear as a merit of its reliability.
Recent advances in observational astrophysics uncovered a bunch of new binary magnetic stars that were considered as rare before. We outline the results of studies of exotic binary and multiple systems with magnetic CP components: HD 6757, HD 34736, and HD 40759.

1 Introduction

Magnetic fields are ubiquitous in space. With the different scale and strength, the fields often take part in birth, life, and death of the stars. Objects with measurable magnetic field appear in all parts of the Hertzsprung–Russell diagram. In her review, L. Ferrario [1] counted different cases of magnetic objects, mostly heavier than two solar masses, and gave an overview of their possible evolutional models. These massive objects have varying age and possess fields ranging from about one gauss to more than thirty kilogauss. In the lower part of the Main Sequence, the dwarfs hold the fields, which are the strongest among the stars with convective envelopes. As an example, recent searches of magnetic fields in M-dwarfs have revealed the surface field of the order of several kG [2]. A weak, the order of gauss field was also found in some evolved objects far beyond the MS. The extensive list of references covering this problem can be found in an introduction to a paper by Mathias et al. [3] devoted to the long-term study of the magnetic field in Betelgeuse.

Magnetic fields of cool stars often are closely related to their activity. Famous solar spots are the most known example of such activity, but it can manifest itself in many other ways. There are cases of radio and X-ray emission attributed to the flares. Chromospheric flares are another way of presentation of the magnetic field. Until recently, it was believed that the activity on the Main Sequence is intrinsic to cool dwarfs and the hottest stars. In contrast, single intermediate-mass stars were considered as relatively quiet objects, even if they host a strong magnetic field. The concept has been changed with a discovery of radio emission

*e-mail: eugene@narit.or.th

from hot B chemically peculiar stars. Now, five Bp-stars with coherent radio emission are known [4].

In the list of magnetic objects, chemically peculiar or CP stars take a special place. They occupy the upper part of the HR diagram in the range of the effective temperature from about 7 000 to almost 25 000 K, and spread over the entire width of the Main Sequence. Three main hypotheses were proposed to explain the origin of stellar magnetic fields. The most evident one is a "turbulent dynamo" hypothesis [5, 6]. This mechanism is responsible for the generation of the magnetic field in stars with strong convection envelopes like our Sun. In the case of CP-stars, dynamo might play an important role in the early stages of stellar evolution. Still, later, during the MS stage, its importance must decrease as the stellar atmospheres become mostly radiative. If the field originated in a deep convective core of a typical Ap/Bp-star, then its emerging time would exceed MS lifetime of the star.

Within the paradigm of the "fossil" field, one can explain a slow decay of the field on the MS. This hypothesis states that the field, which had been originated during the early stage of star formation, during the MS phase slowly dissipates [7, 8]. By default, the observed field takes its origin in the galactic seed field, but the hypothesis does not exclude other sources.

As a combination of the best of two previous hypotheses, the third one has been developed recently by Ferrario et al. [9]. This hypothesis, being concordant to the "fossil" theory, suggests that the seed magnetic field was formed in the process of the active mixing during the stellar merges. In this sense, it could be considered as a partial case of fossil field theory with alternative initial conditions. This scenario must not only explain the low fraction of magnetic CP-stars within the same spectral type but also make clear why is so small the number of close binaries with magnetic components.

Nowadays, there is no consensus in a problem of field origin, but observations mostly witness in favour of fossil nature of the field. At least such fields are observed during the MS stage. In either case, none of the three hypotheses can explain all diversity of the observed parameters of the magnetic CP-stars.

In this regard, binary and multiple stellar systems seem to be the natural test for the proper choice of the origin hypothesis. As the different scenarios predict the contrasting rates of binaries, varying orbital periods, and distinctive individual properties of the stars, it is crucial to study the systems where the components are of the same mass or anomalies.

2 Chemically Peculiar Stars

From the introduction, it becomes clear that in the Upper Main Sequence only chemically peculiar or CP stars could be recognised as magnetic objects. However, CP stars contain many different objects, which form the group of 10–15% of MS stars showing abnormally enhanced or weakened lines of some specific elements in spectra. It was Preston [10] who divided all CP stars into four groups according to a dominating type of anomalies. The A-stars, mostly with deficient Ca and Sc and overabundant heavy metals, fill the first group CP1. The second, CP2 group consists of A-stars with enhanced Fe, Cr, Sr, and some rare-earth elements. In the third group, Preston classified the stars with HgMn type of anomalies. The last group, CP4, contains the stars with abnormally weakened He lines in the spectra. It is worth to note here that the pairs CP1–CP2 and CP3–CP4 share the same range of effective temperatures, but there is the feature that suggests another combination. In total, about 500 stars in groups CP2 and CP4 (and some fraction of the stars with enhanced He lines which are out of this scheme) host strong and stable magnetic fields. Neither example of such field was found for metallic-line Am or HgMn stars. Thus, from lower to higher effective temperature, CP2–CP4 and, respectively, CP1–CP3 form two continuous sequences of magnetic and non-magnetic stars.

The magnetic field of CP2 and CP4 (or hereafter mCP) stars ranges from hundreds of gauss to more than 30 kG, covers the whole surface, and remains stable on a timescale of decades. Measured characteristics of the field vary with a period of stellar axial rotation. Thus, the typical magnetic CP-star looks like a large rotating magnet. This feature differs mCP stars from cool dwarfs, even with a comparable value of their mean surface field. Magnetic Doppler Imaging [11], a unique technique that allows mapping of the surface of these stars, reveals simple field geometry with mostly dipolar structure combined with multipoles of higher orders in some cases [12]. The total number of CP-stars with the confirmed magnetic field is about 500 and continuously grows as the major programmes aimed at searches of such instances are carrying out [13].

Multiplicity is another key difference between the magnetic and non-magnetic groups of CP stars. From numerous studies, it is known that CP1 and CP3 stars tend to appear in binary and multiple stellar systems. The fraction of binaries, in this case, is close to 90% [14]. Many of the stars are in spectroscopic binary systems, but there is also a significant fraction of Am/HgMn in wide pairs [15]. Orbital periods of SB systems containing non-magnetic CP stars fall mostly between 1.2 and 100 days.

In contrast, mCP stars in stellar systems are less frequent. In paper [16] the authors found the rate of binaries with mCP components as low as 43% with apparent lack of magnetic objects in eclipsing binaries. With improvements in observational technique, the number of the known eclipsing magnetic system has grown slightly, but it still far from that for the non-magnetic systems. Avoidance of short-periodic systems is a landmark of magnetic CP stars. Marked differences might be caused by the presence of a strong magnetic field during early evolutional stages and thus they deserve for a detailed study.

The multiplicity of magnetic CP stars has been studied for many years within the individual projects like BinaMIcS [17] and in common surveys like those were carried out in the Special Astrophysical Observatory [18, 19].

3 Magnetic Measurements

The Zeeman effect in spectra is the most reliable way for detection and measure of a stellar magnetic field. In the presence of the field, spectral lines split into polarised components. The split pattern depends on the strength of the field, its orientation, and a configuration of each particular atomic transition. The polarisation of the components depends on field geometry and spatial orientation of the star. The Zeeman splitting in non-polarised spectra is only observed in the stars with a strong field and slow axial rotation. That is why spectropolarimetry is proved to be a powerful tool of weak field detection in rotationally broadened spectra. As a comprehensive introductory course on stellar spectropolarimetry we can recommend the reviews written by D. Landstreet [20] and S. Bagnulo [21].

From spectropolarimetric observations one can possible to measure the mean longitudinal field $\langle B_z \rangle$ which is the line-of-sight component of magnetic field integrated over the visible disc of the star. The distance $\Delta\lambda$ from split circularly-polarised σ-components to the position λ_0 of an undistorted line depends on the longitudinal field $\langle B_z \rangle$ and the Landé factor of the line g_L as follows

$$\Delta\lambda = \pm 4.67 \cdot 10^{-13}\, g_L\, \lambda_0^2 \langle B_z \rangle.$$

Wavelength here is measured in angstroms, the mean longitudinal field is in gauss.

If the magnetically split components are visible in unpolarized spectra, one can measure the mean surface magnetic field strength $\langle B \rangle$. This parameter is defined as a disc-integrated modulus of the field strength.

Practically, the Zeeman splitting rarely looks like a simple triplet. For most of the lines, π and σ-components have a complex asymmetric structure that complicates their measuring.

4 Three Examples of Multiple Systems

In this section, we present three magnetic CP stars in binary and multiple systems the study of which may help to shed light on the origin, evolution, and other aspects of the stellar magnetism. The magnetic field of these stars was found with the use of the Russian 6-m telescope BTA.

4.1 HD 6757

The magnetic field of HD 6757 was measured for the first time by Kudryavtsev et al. [22]. As the star shows a constantly positive field which varies within a small range, for a long time it has been considered as a slow rotator. Research of more than 15 individual measures of the longitudinal magnetic field does not detect any unambiguous periodicities.

In high-resolution echelle spectra of the star, the line of Fe II 6149Å, which is a standard tool of the magnetic field diagnostic in unpolarized data [23], displays the Zeeman splitting corresponding to $\langle B \rangle$ = 9.9 kG. The Nd III 6145Å line with a triplet-like Zeeman pattern suggests $\langle B \rangle$ closer to 11 kG (Fig. 1). Like the longitudinal field, the mean modulus does not change significantly, which is typical for the slowly-rotating and pole-on stars.

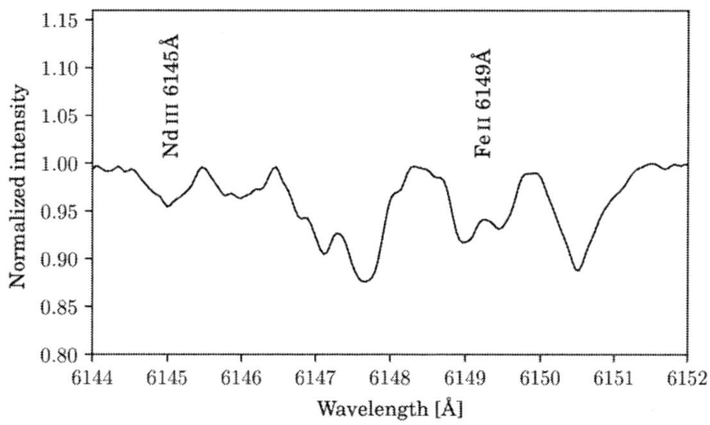

Figure 1. A portion of an echelle spectrum of HD 6757 showing the Zeeman splitting caused by the magnetic field.

We have extracted available TESS[1] photometry to check for periodic variability. At first glance, the TESS lightcurve contains two signals with periods 9.205 and 1.80098 days (Fig. 2), but the fine detrending is required. Until then, it is impossible to distinguish the sources of the signals.

HD 6757 is a well-known member of a visually binary system ADS 936. At the same time, the echelle spectra of this star obtained in 2006 and 2008 display the trace of at least one more star in hydrogen lines (Fig. 3). The number of available high-resolution spectra is still insufficient to estimate the orbit of this system. The measured radial velocity of the weak spectroscopic component (RV = 40–50 km s^{-1} in the red wing of H$_\beta$ line) does not exclude a short orbital period of the spectroscopic system. Thus, we expect that the 9-days period may reflect the orbital motion in the inner spectroscopic system.

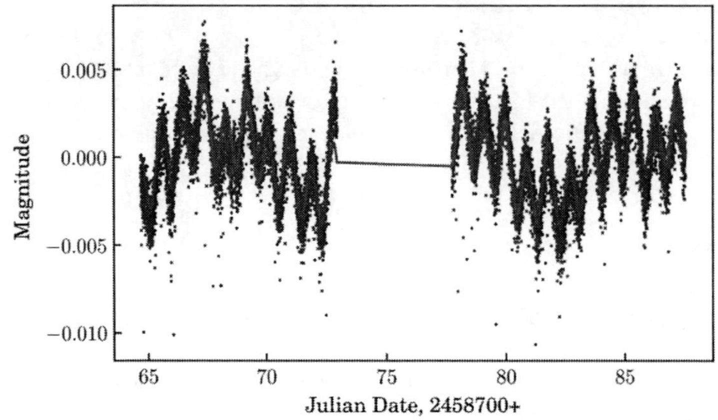

Figure 2. TESS photometry of HD 6757 and its fit with a combination of two periodic functions.

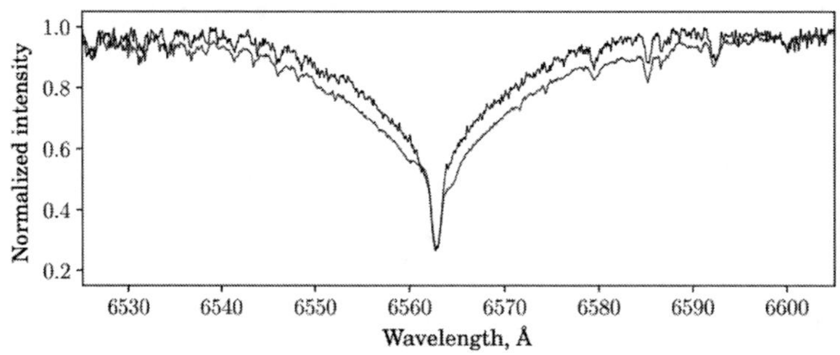

Figure 3. Hydrogen line H_β in HD 6757 spectra. The deformation of the transition between the core and wings of the line is clearly seen.

The shorter period could arise from the axial rotation of the CP component. Given $7\,\mathrm{km\,s^{-1}}$ as a projected rotational velocity measured from the spectra, and $R = 2R_\odot$ as the most probable radius corresponding to the spectral type, from the oblique rotator model [24] we can expect the dipolar magnetic field with $B_\mathrm{p} \approx 13\,\mathrm{kG}$, in good agreement with observations.

In the future, we plan to decompose this multiple-star system into components and estimate their physical parameters. That will help to understand how the multiple hierarchical systems evolve.

4.2 HD 34736

This 7.82 mag CP star with enhanced lines of silicon in a spectrum was recognized as a magnetic star in 2013 [25]. In observations, the longitudinal field varies from -6 to +6 kG with a period of 1.2799 days (Fig. 4). This value corresponds to the rotational period of HD 34736,

[1]This paper includes data collected by the TESS mission. Funding for the TESS mission is provided by the NASA Explorer Program.

which dominates in TESS photometry (Fig. 5). The original lightcurve was decomposed by Z. Mikulášek onto two components. The first extracted curve has a complex shape typical for CP stars. The second curve also has a complex shape, but its period is much shorter — only 0.52 days. The origin of the second curve is still unclear because it might reflect both extremely fast axial rotation of a companion star and pulsation of either component.

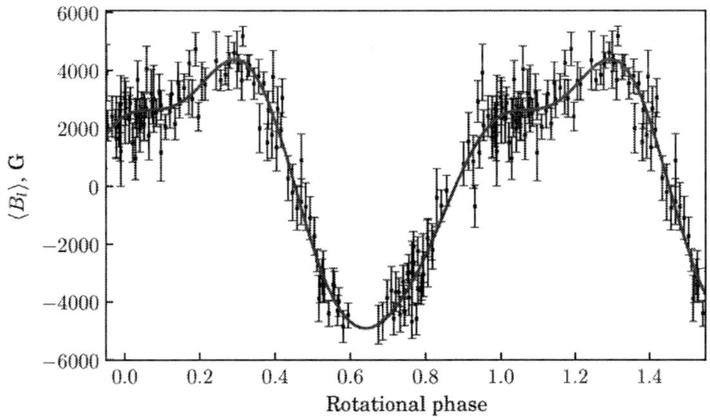

Figure 4. The longitudinal magnetic field of HD 34736 phased with its rotational period.

HD 34736 is a spectroscopic binary system consisting of a magnetic star with relatively narrow lines in the spectrum and a broad-lined star without a measurable magnetic field. Both components are of late B type. With this respect, it is important to establish the real source of the fast periodicity. In the case of fast-rotating secondary star, its variability may suggest the spotted surface and consequently abnormal chemical composition. In this respect, the secondary star by its properties might resemble the well-known mCP variable CU Vir [26]. Alternatively, the spectral types of both components cannot exclude the pulsation of SPB-type like in the wide binary system *o* Lup [27]. Either scenario looks possible and is essential for understanding the evolution of binary systems.

All the results of the current study will appear in a dedicated paper.

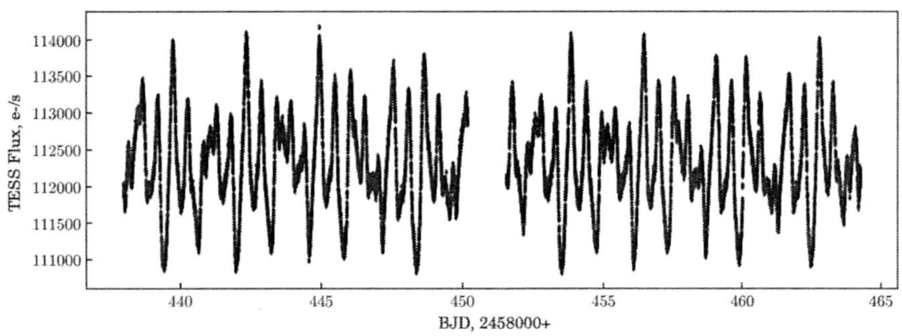

Figure 5. TESS photometry of HD 34736.

4.3 HD 40759

Like in the case of HD 6757, the magnetic field of HD 40759 was discovered by Kudryavtsev et al. [22]. Later this star was included in a spectropolarimetric survey of stars in the Orion OB1 association as HD 40759 is a known member of its subgroup *c* [28]. During the full period of observation, we have collected more than ten individual measurements of $\langle B_l \rangle$, although the period of its variability remains unclear.

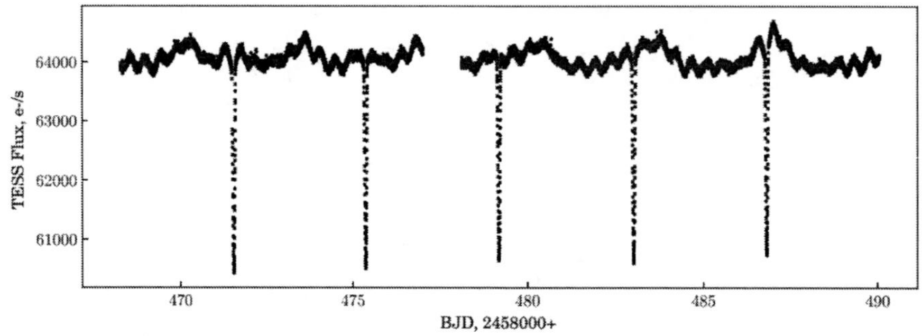

Figure 6. TESS photometry of HD 40759.

The lightcurve of TESS photometry has a waving shape with explicitly seen eclipses (Fig. 6). From an empiric analysis of photometric series, we have derived at least four periodic signals. The system does not exhibit secondary eclipses, and the time between the deep minima (eclipses) is 3.815 days. The second-longest period of 3.432 days may reflect the rotation of the magnetic component, but it cannot fit the measured field $\langle B_l \rangle$. The two other periods are 0.392 and 0.461 days. The origin of the short periodic signals is unclear. It is unlikely that this type of variability comes from rotation. Instead, it might be expected in the case of pulsation. This hypothesis finds support from the fact of stellar multiplicity. HD 40759 is a spectroscopically triple SB3 system as it follows from spectra, obtained with the Russian 6-m telescope. From the measured radial velocities of weaker components and assuming the orbital period equals to time between eclipses, we have tried to reconstruct a spectroscopic orbit of the system. Because of the small number of observations with resolved components, at this point we can find only a set of possible solutions which suggest the elongated orbit with $e = 0.4$–0.5 and minimal projected masses $M \sin^3 i$ of the components around $2M_\odot$ (Fig. 7). Such combination lets us suppose that the fainter stars are either a pair of two δ Sct or SPB-type pulsators, depending on the inclination of the orbit. The magnetic CP star is orbiting the inner pair with a very long and undefined period.

5 Summary

The three examples from this paper show that the occurrence of magnetic CP stars in binary and multiple systems might be higher than it was commonly believed. Concerning spectroscopically binary stars, one needs to accept that the observational selection is underestimated, first of all, due to the low number of corresponding observations per one system. The examples of HD 6757 and HD 34736 perfectly demonstrate the value of a large number of high-resolution spectra. Both stars do not expose the lines of all components most of the time.

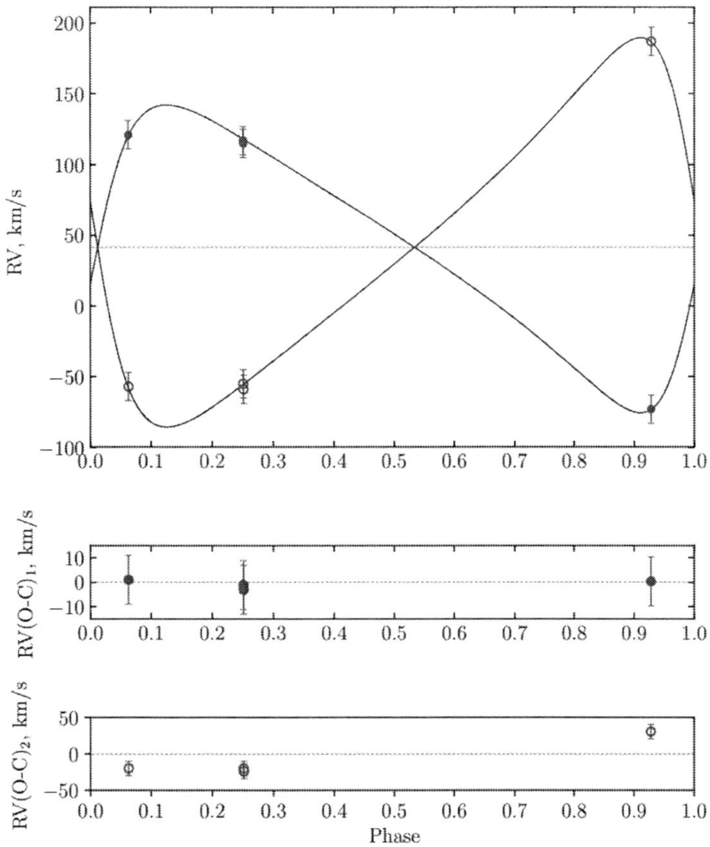

Figure 7. Possible orbital solution for HD 40759.

In searches of new interesting binary systems, good results can be achieved when the high-resolution spectroscopy combines with photometry. Time-resolved data obtained by the TESS satellite suit perfectly for such purposes. Solely from photometry one can find the pulsating components in the systems like HD 34736 or HD 40759 which are too faint for time-resolved and high-resolution spectroscopy. The fact of the presence of pulsating stars in the system is especially important because depending on the amplitude and period, it allows putting the restrictions on expected physical parameters of weak companions. HD 40759, in this sense, appears as a unique laboratory.

All the results presented in this paper are preliminary. For the stars HD 6757 and HD 40759 in an observational season 2019/2020 with the 2.4-m Thai National Telescope we have collected a substantial set of echelle spectra which would be included in the research. After all, the results will be published in the relevant papers.

Acknowledgements. The author is thankful to his colleagues I. Romanyuk, I. Yakunin, A. Moiseeva (all from the Special Astrophysical Observatory, Russia), Z. Mikulášek (Masaryk University, Czech Republic), and D. Mkrtichian (NARIT, Thailand) for their valuable help in collecting and analysis of observational data.

References

[1] L. Ferrario, Contributions of the Astronomical Observatory Skalnate Pleso **48**, 15 (2018), 1801.09368
[2] D. Shulyak, A. Reiners, E. Nagel, L. Tal-Or, J.A. Caballero, M. Zechmeister, V.J.S. Béjar, M. Cortés-Contreras, E.L. Martin, A. Kaminski et al., Astronomy and Astrophysics **626**, A86 (2019), 1904.12762
[3] P. Mathias, M. Aurière, A. López Ariste, P. Petit, B. Tessore, E. Josselin, A. Lèbre, J. Morin, G. Wade, F. Herpin et al., Astronomy and Astrophysics **615**, A116 (2018), 1804.01831
[4] B. Das, P. Chandra, M.E. Shultz, G.A. Wade, Monthly Notices of the RAS **489**, L102 (2019), 1908.09110
[5] L. Mestel, J.D. Landstreet, *Stellar Magnetic Fields* (2005), Vol. 664, p. 183
[6] L. Mestel, D. Moss, Monthly Notices of the RAS **405**, 1845 (2010)
[7] E.F. Borra, J.D. Landstreet, L. Mestel, Annual Review of Astron and Astrophys **20**, 191 (1982)
[8] C. Neiner, S. Mathis, E. Alecian, C. Emeriau, J. Grunhut, BinaMIcS, MiMeS Collaborations, *The origin of magnetic fields in hot stars*, in *Polarimetry*, edited by K.N. Nagendra, S. Bagnulo, R. Centeno, M. Jesús Martínez González (2015), Vol. 305 of *IAU Symposium*, pp. 61–66, 1502.00226
[9] L. Ferrario, J.E. Pringle, C.A. Tout, D.T. Wickramasinghe, Monthly Notices of the RAS **400**, L71 (2009)
[10] G.W. Preston, Annual Review of Astron and Astrophys **12**, 257 (1974)
[11] N. Piskunov, O. Kochukhov, Astronomy and Astrophysics **381**, 736 (2002)
[12] O. Kochukhov, arXiv e-prints arXiv:1912.07241 (2019), 1912.07241
[13] V.D. Bychkov, L.V. Bychkova, J. Madej, Monthly Notices of the RAS **394**, 1338 (2009)
[14] S. Hubrig, J. González, M. Schöller, *Multiplicity of Chemically Peculiar Stars* (2010), Vol. 435 of *Astronomical Society of the Pacific Conference Series*, p. 257
[15] M. Schöller, S. Correia, S. Hubrig, N. Ageorges, Astronomy and Astrophysics **522**, A85 (2010), 1010.3643
[16] F. Carrier, P. North, S. Udry, J. Babel, Astronomy and Astrophysics **394**, 151 (2002), astro-ph/0208082
[17] E. Alecian, C. Neiner, G.A. Wade, S. Mathis, D. Bohlender, D. Cébron, C. Folsom, J. Grunhut, J.B. Le Bouquin, V. Petit et al., *The BinaMIcS project: understanding the origin of magnetic fields in massive stars through close binary systems*, in *New Windows on Massive Stars*, edited by G. Meynet, C. Georgy, J. Groh, P. Stee (2015), Vol. 307 of *IAU Symposium*, pp. 330–335, 1409.1094
[18] Y.Y. Balega, V.V. Dyachenko, A.F. Maksimov, E.V. Malogolovets, D.A. Rastegaev, I.I. Romanyuk, Astrophysical Bulletin **67**, 44 (2012)
[19] E.A. Semenko, I.I. Romanyuk, I.A. Yakunin, A.V. Moiseeva, D.O. Kudryavtsev, *Binary Stars in the Orion OB1 Association, Subgroup a* (2019), Vol. 518 of *Astronomical Society of the Pacific Conference Series*, p. 31
[20] J.D. Landstreet, *Basics of spectropolarimetry*, in *New Windows on Massive Stars*, edited by G. Meynet, C. Georgy, J. Groh, P. Stee (2015), Vol. 307 of *IAU Symposium*, pp. 311–320
[21] M.I. Mishchenko, Y.S. Yatskiv, V.K. Rosenbush, G. Videen, *Polarimetric Detection, Characterization and Remote Sensing* (2011)

[22] D.O. Kudryavtsev, I.I. Romanyuk, V.G. Elkin, E. Paunzen, Monthly Notices of the RAS **372**, 1804 (2006)
[23] G. Mathys, S. Hubrig, J.D. Landstreet, T. Lanz, J. Manfroid, Astronomy and Astrophysics Supplement **123**, 353 (1997)
[24] J.J. Monaghan, Monthly Notices of the RAS **163**, 423 (1973)
[25] E.A. Semenko, I.I. Romanyuk, D.O. Kudryavtsev, I.A. Yakunin, Astrophysical Bulletin **69**, 191 (2014), 1404.5937
[26] R. Kuschnig, T.A. Ryabchikova, N.E. Piskunov, W.W. Weiss, M.J. Gelbmann, Astronomy and Astrophysics **348**, 924 (1999)
[27] B. Buysschaert, C. Neiner, A.J. Martin, M.E. Oksala, C. Aerts, A. Tkachenko, E. Alecian, MiMeS Collaboration, Astronomy and Astrophysics **622**, A67 (2019)
[28] I.I. Romanyuk, E.A. Semenko, I.A. Yakunin, D.O. Kudryavtsev, Astrophysical Bulletin **68**, 300 (2013)

Predictive and Characterized Pulsar Evolution From its Optical Pulsation

Rizchel Masong[1,2*]

[1]De La Salle University Manila, College of Science, Physics Department, 2401, Taft avenue, Manila, Philippines.
[2]De La Salle Araneta University, College of Arts and Sciences, Science Department, Victoneta Avenue, Malabon, Philippines

Abstract. The measured and predicted variables of pulsar PSR B1937+21 were explored from the energetics obtained from the previous observation made. The target parameters such as the brightness, luminosities, and optical efficiency. A series of observations were analysed to determine the individual values of the parameters chosen. Within the extended area of optical pulsation, the values were plotted to determine the relation of it to synchrotron emission. The results were presented graphically to see whether PSR B1937+21 will remain consistent with the expected pulsar model. The researcher suggests that simulation and correlation of previous observation made together with a new one to explore the cause of optical pulsation from mass transfer and spectral evolution.

Introduction

Pulsar is known as is a highly magnetized rotating neutron star that emits beams of electromagnetic radiation out of its magnetic poles and originally know to be observed under radio waves. However, there is a surprising event that happened in PSR J1023+0038. There was a state change that occurred in PSR J1023. There was a disappearance of radio pulsations and a development of accretion disk (Stappers, Patruno, Tendulkar, 2014) in a simultaneous observation and study. Further strengthening the idea that most of the optical and X-ray luminosity comes from an accretion process and not from an intrabinary shock or from the spin-down luminosity of a rotation-powered pulsar (Patruno, 2014). Since, we saw that radio pulsation disappears as accretion disk develops. What type of pulsations can we now detect? X-ray/optical correlation found in NS LMXBs with the addition of J1023. The correlation exists between the reprocessed light and the X-ray emission coming from the accretion process (Patruno, et al. 2014). A pulsar is also observable under the optical, Xray, and UV regime (Ambrosino, et al. 2018) on similar pulsar PSR J1023. Optical pulsed emission could be due to synchrotron emission by relativistic electrons in the magnetosphere of a rotation-powered pulsar. This also demonstrates that the magnetosphere of old, weakly magnetic and

* Corresponding author: rizchel_masong@dlsu.edu.ph

quickly spinning neutron stars can give rise to such signals when surrounded by an accretion disk

PSR B1937+21 (PSR J1939+2134) on the other hand, is a pulsar located in the constellation Vulpecula and was the first millisecond (MS) pulsar to be discovered (Backer et al. 1982), it has a period of 1.56 ms, and was the fastest-spinning known pulsar. The profile of PSR B1937+21 consists of a main pulse, which has at least two components spaced close to each other, and an interpulse and the separation angle. $\Delta\varphi_{MP-IP} = 174°$ remains constant for a wide frequency range, from 0.32 to 2.38 GHz (Stinebring & Cordes 1983). The constant $\Delta\varphi_{MP-IP}$ could be perceived as an indication that the emission originates from two different magnetic poles. This view is further supported by the lack of low-level emission for most of the period, at least with the current observations (e.g., Camilo 1995). It came as a surprise to find qualitatively similar behavior from PSR B1937+21 that shows frequent, very strong radio pulses extending to hundreds of times the mean pulse intensity. the first (and still fastest known) millisecond pulsar.

The Radhakrishnan & Cooke (1969) explanation of the Monotonic sweep of linear polarization across the Vela pulsar's average pulse profile in terms of this polar cap model provided the cornerstone upon which most subsequent work has been based (Sallmen & Backer 1995; Cognard et al. 1996). When a pulsar rotates fast enough the separation between tangents to the outermost field lines which connect to the interstellar medium spreads in proportion to $P^{-0.5}$ for a simple dipole geometry when measured at the same altitude. If pulsar radiation fills a fixed fraction of this 'open' field zone, then we could expect a similar scaling of observed pulse width, measured in angle not time, with period. A dependency of emission altitude on period can, of course, alter this relation. Usual analysis and studies where made in relating width-period relation since the early days of pulsar discovery. The technique demonstrated the dependence of pulsar property in objects within the same line of sight, dipole axis, and spin axis geometries were observed and compared (Lyne & Manchester 1988). What arises from these studies is that the central component of the beam provided by pulsar rotation has a steep spectrum. And also displays sign reversal of circular polarization.

PSR B1937+21 (also referred to as PSR J1939+2134) was the first MSP discovered (Backer et al. 1982) and for 25 years, the fastest spinning pulsar known (P = 1.55 ms). With a spindown luminosity of $\dot{E} = 1.1 \text{ x } 10^{36} erg\, s^{-1}$, it is the second most energetic MSP detected. This isolated pulsar lies in the Galactic plane at a distance of ~3.5 kpc. At X-ray energies, PSR B1937+21 has been detected as a pulsed source with ASCA (Takahashi et al. 2001), BeppoSAX (Nicastro et al. 2004), RXTE (Guillemot et al. 2012), Chandra, and XMM-Newton (Ng et al. 2014). In hard X-rays, the pulsar has so far only been detected up to ≈13 keV with RXTE (Guillemot et al. 2012). Like PSR B1821-24, this MSP shows sharp pulsations with a small duty cycle and nearly 100% pulsed fraction. In this paper, we present an analysis of the optical properties of PSR B1937+21 in terms of the observation made by NuSTAR, Chandra, and XMM-Newton data, and archived in ATNF Pulsar Catalogue made by CSIRO (Manchester, R. N., Hobbs, G.B., Teoh, A. & Hobbs, 2005)

Observations and Data Analysis

The observations used were made using the Large European Array for Pulsars (LEAP); a tied-array telescope comprised of various combinations of five European 100-m class radio telescopes: the Effelsberg Telescope, the Lovell Telescope, the Nancay Radio Telescope

(NRT), the Westerbork Synthesis Radio Telescope (WSRT), and the Sardinia Radio Telescope (SRT). Observations with the participating telescopes are made simultaneously and are combined coherently into a tied-array beam, forming a virtual 195-m dish when using all five telescopes (McKee et. al, 2018).

Table 1. Summary of the LEAP observation of PSR B1937+21.

Date	MJD	Telescopes	f_c (MHz)	BW (MHz)	T_{obs} (sec)	$S/N_{profile}$	Coherency	N_{MGP}	N_{IGP}	N_{GP}
2012-09-23 (1)	56193	EJW	1412	64	1240	719	80%	125	77	202
2012-09-23 (2)	56193	EJW	1412	64	1020	808	86%	124	74	198
2013-07-27	56500	EJW	1404	112	1800	525	63%	138	106	244
2013-08-25	56529	EW	1364	64	2700	1065	98%	278	154	432
2014-05-22	56799	ESW	1396	128	2760	904	86%	168	83	251
2014-07-07	56845	EW	1420	80	2770	435	88%	172	84	256
2014-07-27	56865	EW	1380	96	2760	492	97%	31	22	53
2014-08-24	56893	EJNW	1396	128	2490	852	92%	155	75	230
2014-10-15	56945	EJNW	1380	64	1470	587	57%	139	76	215
2015-02-25	57078	EJNW	1396	128	2120	715	76%	145	94	239
2015-03-26	57107	EJNSW	1396	128	2510	1232	82%	171	87	258
2015-04-17	57129	EJNSW	1396	128	2510	1202	80%	137	52	189
2015-06-20	57193	EJW	1396	128	2230	598	100%	112	63	175
2015-07-19	57222	EJS	1412	96	2790	458	67%	112	92	204
2015-09-18	57283	EJN	1404	112	2520	429	78%	90	32	122
2015-10-10	57305	EJ	1412	96	2790	543	86%	112	52	164
2015-11-07	57333	EJS	1396	128	2790	729	99%	81	55	136
2015-12-12	57368	EJNS	1396	128	2520	755	75%	138	55	193
2016-01-09	57396	EJNS	1396	128	2520	1122	83%	154	99	253
2016-02-05	57423	EJNS	1404	112	2390	629	75%	97	49	146
2016-04-07	57485	EJNS	1388	80	2380	924	78%	110	48	158
Mean	-	-	-	-	2337	709	82%	133	70	203
Total	-	-	-	-	49080	-	-	2789	1476	4265

Monthly observations have been made in LEAP mode since February 2012, and is presented in Table 1. It consists of the telescope used in the observation proper, the frequency and bandwidth, and the duration of the observations made. There is a difference in bandwidth and either caused by instrumental setup or due to data loss. Also, the ephemeris of PSR B19387+21 and derived parameters. Obtained from observations with the Effelsberg 100-m telescope and Arecibo archival data. Figures in parentheses represent uncertainties (2σ) in the last digits quoted.

Table 2. Summary of the LEAP observation of PSR B1937+21.

Parameter	**Value**
Right Ascension (J2000)	$19^h\ 39^m\ 38^s.5600084(8)$
Declination (J2000)	$21°\ 34'\ 59''.13548(14)$
Frequency (f, Hz)	641.928246349481(9)
Freq. derivative ($\dot{f}$, Hz s^{-1})	$-4.330999(2) \times 10^{-14}$
Freq. 2 nd deriv. ($\ddot{f}$, Hz s^{-2})	$1.558(8) \times 10^{-26}$
DM[a] (cm^{-3} pc)	71.03998(6)
Epoch of ephemeris (MJD)	52328.0
Range of validity (MJD)	45986 – 52764
RA proper motion (mas yr^{-1})	−0.128(8)
Dec proper motion (mas yr^{-1})	−0.486(12)
Gal latitude (l)	57.5215
Gal longitude (b)	−0.2697
Parallax (π, mas)	< 0.28
Distance[b] (d – from DM, kpc)	3.6
Surface magnetic field (B_S, G)	4.1×10^8
Light-cylinder B (B_L, G)	1.0×10^6
Spin-down energy loss ($\dot{E}$, erg s^{-1})	1.1×10^{36}
$\dot{E}/4\pi d^2$ (erg cm^{-2} s^{-1})	7.1×10^{-10}

Exploring the temporal analysis made (Cusamano, et. al, 2003), showed that the folded light curve of the MECS data together with the radio one obtained with the 100- m Effelsberg radio-telescope of the LEAP network. On the other hand, Figure 2 shows the X-ray pulse profile in 3 energy ranges. Non-detection of the interpulse in the ASCA data is explained by their lower statistical significance (227 total photons in 1.7–6.5 keV).

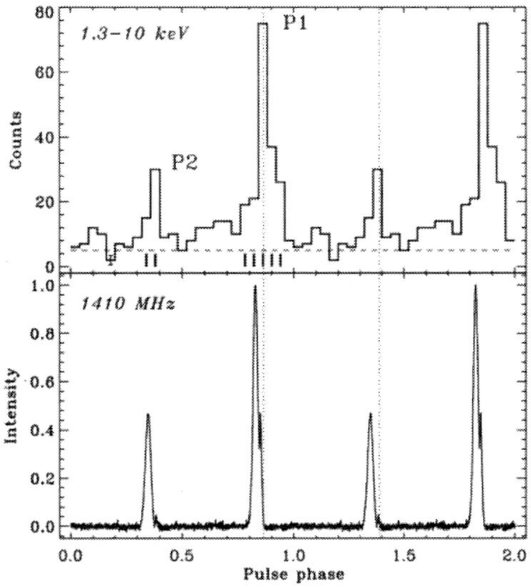

Figure 1. Radio pulse profile and the 1.3–10 keV BeppoSAX X-ray profile phase aligned with the radio giant pulses.

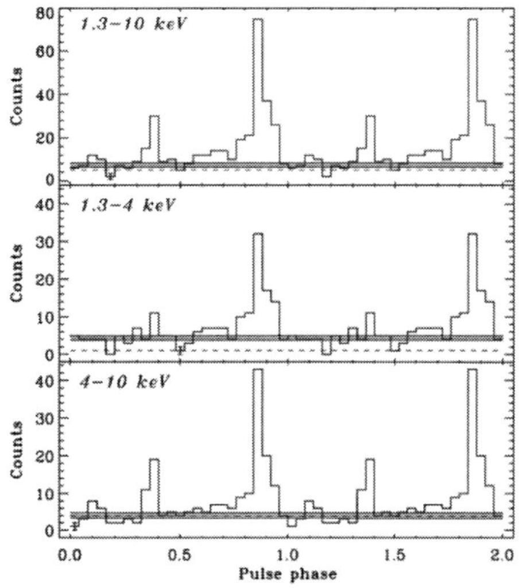

Figure 2. The X-ray pulse profile in the full MECS X-ray band 1.3–10 keV and two sub-bands.

From the observations made. Exploring the optical property of pulsar in terms of its brightness, spin down luminosities, and optical efficiency made it possible. For giant pulses of pulsars, the energy density is comparable with that of oscillation in the inner gap. It means that we obtain radiation at the giant pulsar instants directly from the gap by-passing the magnetosphere plasma. This explains that pulsars as an emitter has a main characteristic which known as the brightness temperature and spectrum as well as polarization that can be considered as grounds for analysis (Kontorovich, 2015). The brightness b is defined by equation (1).

$$b = \frac{L}{4\pi d^2} \tag{1}$$

Based on this, we can view the main emission properties of millisecond pulsar which show non-thermal emission only, in order to account for the contribution to the flux by the field source listed by Becker et al. (2003). We can also explore the spin down rate based on the given logarithm of luminosity and logarithm of spin down rate.

As a pulsar emits radiation, the energy is extracted from its rotation. By observing the luminosity of a pulsar's nebula, we can estimate a rate at which pulsar is slowing down. This is known as the spin-down luminosity which is defined by the equation (2) which is the ratio from the $-\dot{E}$.

$$-\dot{E} = \frac{4\pi^2 I \dot{P}}{P^3} \tag{2}$$

Optical efficiency on the other hand is the ratio of the luminosity and it's spin down power luminosity. Since the luminosity can be obtain from the solution of equation 2, we can obtain its ratio with the spin-down luminosity. The value can be achieved by the equation 3.

$$\eta_{opt} = \frac{L_x}{\dot{E_{sd}}} \tag{3}$$

Based on the mentioned parameters together with the observations made. Exploration about the pulsar property will be interpreted.

Results

Since the energetics was obtained from the previous observations made for PSR B1937+21, luminosity can be easily measure and interpret. We are expecting a value of brightness that is equivalent to 3.6 x 10^6 Wm^{-2}. And by plotting the optical luminosities $Log\, L_x$, and $Log\, \dot{E}$ showed in figure 3 shows the main emission properties of millisecond pulsars and demonstrates non-thermal emission. And in order to account for the contribution to the flux by the field source listed by Becker et al. (2003). We have $Log\, \dot{E}$ to be equivalent to 36.0 erg/s. While $Log\, L_x$ to be equivalent to 31.57 erg/s and displays that pulsar PSR B1937+21 which exhibits a property of being a dipole and has a maximum luminosity for synchrotron emission.

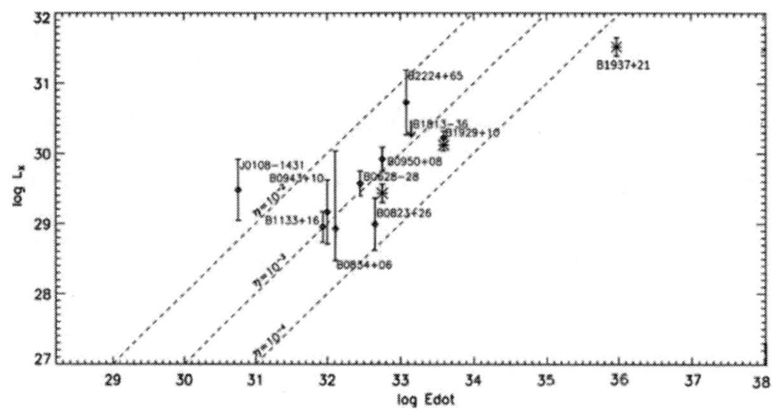

Figure 3. Plot of logarithm of spin down luminosity and logarithm of optical luminosity with pulsars in both radio and optical regime.

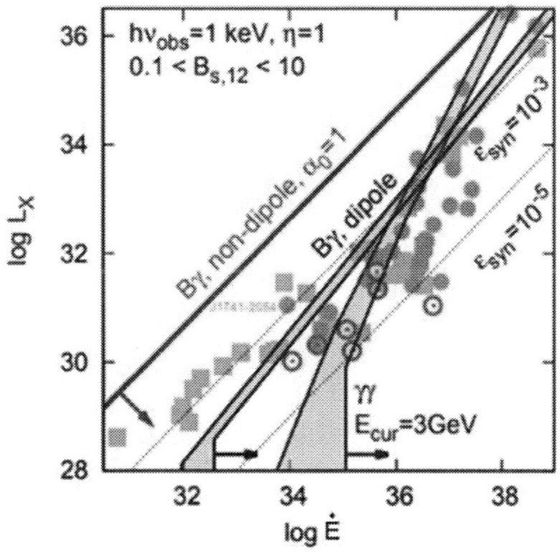

Figure 4. Maximum luminosities of synchrotron radiation in 1 keV (upper) and 1 eV (lower) for two-photon collision case (γγ, blue), magnetic pair creation cases (Bγ) with dipole field (black) and non-dipole fields (red) as a function of the spin-down luminosity L_{sd}.

Comparing the plot of Figure 3 to Figure 4 we can see that PSR B1937+21. We can observe that pulsar PSR B1937+21 exhibits a property of being a dipole and has a maximum luminosity for synchrotron emission which confirms the pulsar model. This confirms that PSR B1937+21 has a projected optical pulsation in terms of synchrotron emission obtained from the optical luminosity and spin down luminosity. We also consider the range of the surface magnetic field 0.1 < Bs,12 < 10 G. Its dependence of the luminosity on the magnetic field is described by $L_{syn} \propto B - 1/4$ s. (Kisaka and Tanaka, 2017). Exploring further the spin down luminosity and comparing it with the sun and crab pulsar. We can see that the pulsations/pulse luminosity is close by a factor of 10^2 L⊙ which can be seen in Table 3. From this we can agree that

Table 3. Comparison of the PSR B1937+21 and Crab Pulsar.

Pulsar	**Period**	**Moment of Inertia (I)**	$\dot{P}$	**Spin Down Luminosity**
PSR B1937+21	0.00156 s	10^{45} g/cm^3	10^{-19}	10^{36} ergs/s or 3.82 x 10^3 L⊙
Crab Pulsar	0.033 s	10^{45} g/cm^3	$10^{-12.4}$	4 x 10^{38} ergs/s or 10^5 L⊙

The spin down luminosity of PSR B1937+21 is relatively closer to the value of Crab pulsar by a factor of 10^2 $L_{\odot}$. Further comparison of apparent brightness is advisable to check the interpretation of differences in luminosity. The luminosity observed by PSR B1937+21 shows it compatibility with synchrotron emission of a standard pulsar model. Lastly, optical luminosity together with the spin down power shows that PSR B1937+21 doesn't fall on a radio-active regime as shown in Figure 3.

It advisable to explore the spectral evolution and mass transfer of the pulsar by performing a direct observation and time series analysis. Comparison of the observation records and a new one will verify the correlation of the target parameters. Simulations and actual observation are recommended to test the evolution of pulsar as we analyse the optical properties of it.

My deepest gratitude comes first to my institute for funding my presentation. And to my research adviser, colleagues, and friends for sharing some insights. Especially, to my partner, thank you very much for understanding.

References

1. F. Ambrosino, A. Papitto, L. Stella, F. Meddi, P. Cretaro, L. Burderi, T. Di Salvo, G. L. Israel, A. Ghedina, L. Di Fabrizio, L. Riverol, *Nature Astronomy* (2018)
2. D.C. Backer, *Astrophys. Astr.* (1995)
3. G. Cusumano, T. Mineo, E. Massaro, et al. *Astronomy and Astrophysics* (2003)
4. S. Kisaka, and S. J. Tanaka, *IOP Conf. Series: Journal of Physics: Conf. Series* **932** (2017)
5. V.M. Kontorovich, *Astronomy Reports* volume **59**, pages 277–287 (2015)
6. J. W. McKee, B. W. Stappers, C. G. Bassa, S. Chen, I. Cognard, M. Gaikwad, G. H. Janssen, R. Karuppusamy 2, M. Kramer, K. J. Lee, K. Liu, D. Perrodin, S. A. Sanidas, R. Smits, L. Wang, and W. W. Zhu, *Nature Astronomy*, 1–21 (2018)
7. L. Nicastro, G. Cusumano, O. L¨ohmer, M. Kramer, L. Kuiper, W. Hermsen, T. Mineo, and W. Becker, *Astronomy and Astrophysics* (2004)
8. A. Patruno, S. Bogdanov, A. M. Archibald, C. Bassa, J. W. T. Hessels, G. H. Janssen, B. W. Stappers, arXiv:1402.6324 [astro-ph.HE] *Astrophysical Journal* (2014)
9. B. W. Stappers, A. M. Archibald, J. W. T. Hessels, C. G. Bassa, S. Bogdanov, G. H. Janssen, V. M. Kaspi, A. G. Lyne, A. Patruno, S. Tendulkar, A. B. Hill, T. Glanzman, *Nature Astronomy*, (2014)
10. M. Takahashi, S. Shibata, K. Torii, Y. Saito, N. Kawai, M. Hirayama, T. Dotani, D. Gunji, H. Sakurai, I. H. Stairs , and R. N. Manchester, *The Astrophysical Journal*, **554**:316-321 (2001)

On Multiparticle Production in Very High Energy Scattering

Pulkit Agarwal[1,*], *Han Wei* Ang[1], *Zongjin* Ong[1], *Aik Hui* Chan[1] and *Choo Hiap* Oh[1]

[1]Department of Physics, Faculty of Science, 2 Science Drive 3, Singapore

Abstract. A phenomenological model of particle production and hadronisation in high energy collisions is formulated using Dirac fields in Yukawa-like interaction and the resulting stochastic equation is solved numerically. Different initial conditions are used to compare particle-particle (ψ ψ) and particle-antiparticle (ψ^* ψ) interactions. It is shown that in this simplified view, there is a clear difference between the final multiplicity distributions resulting from the two initial conditions. To model the restricted phase space (limited pseudorapidity) measurements in experiment, a "loss" function is also proposed to account for the undetected particles close to the beam line.

1 Introduction

Particle production is a high energy phenomena that stems from the Einstein's relativistic physics. It is highly relevant in the domain of quantum systems interacting at energies that are significantly larger than the mass scales of known matter. Experiments done at the LHC and all previous collaborations aim to study matter at smaller and smaller length scales with each generation to probe the physics of the early universe. Particle production and hadronisation is an important aspect of what happened after the big bang and our understanding of it is incomplete. Standard model physics is perturbative by design and we do not have a complete map of what goes on in the large coupling sector of Quantum Chromodynamics (QCD), which is the relevant theory for particle production.

Our understanding of QCD in terms of perturbation theory is limited due to the running nature of the coupling constant. Studies of branching models circumvent this problem in an interesting way. Branching models derived from perturbation theories result in stochastic equations of evolution that are independent of the coupling. It is therefore a valid assumption that these equations are still applicable when the coupling is strong [1]. Over the years, such branching models have proved to be a helpful guide to studying branching processes in high energy experiments[†]. This is one of the non-perturbative approaches employed to model physics of the soft sector. The study of hadronic multiplicity at high energies is a highly contested topic as there is no first principles explanation from QCD to the observed data. This further strengthens the case for stochastic phenomenological studies

* Corresponding author: p.agarwal@u.nus.edu

† A concise discussion on various branching models and their properties can be found in [1].

that take into consideration the relevant physics while making certain simplifying assumptions to deal with the problem at hand.

The most popular parton branching equation was first discussed by Giovannini [2]. His branching model has been extensively studied in the final decades of the last century [3-17]. Such branching models work under the following assumptions: particle production occurs in three distinct steps. First, partons showers are created through the branching process birthing from a bunch of highly virtual partons. Next, these partons hadronise through an unknown process. Finally, these hadrons are detected in our experiment subject to detector design and efficiency. This final step is best handled through combinatorics. Unfortunately, the process of hadronisation is still not very well understood.

In this work, we propose a stochastic branching model based on these principles with the interaction term $\lambda\ \psi^*\ \varphi\ \psi$. This is similar to the effective nuclear force described by Yukawa interaction and an exchange of pions. Although Yukawa processes have been studied in quite some detail in the literature, as a branching process they are mostly studied as a φ^3 process [18]. This is one of the more successful models that has the Furry distribution as its solution [19,20]. However, since this is a scalar theory, there is no distinction between particles and antiparticles. It is therefore impossible to probe the difference, if any, between particle-particle and particle-antiparticle interactions [21]. Giovannini's parton branching equation also fails at this count. It does not make any distinction between quark and anti-quark branching probabilities. We shall address these shortcomings in this work and showcase how this distinction between particle-particle and particle-antiparticle interactions manifests itself in our model.

2 Stochastic Branching Models

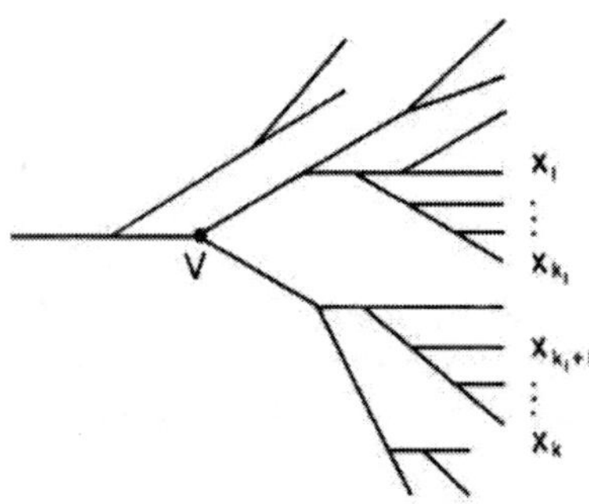

Fig. 1. A schematic diagram to represent a typical production tree‡.

Our setup aims to model the difference between *pp* and $\bar{p}p$ collisions [21] which Giovannini's parton branching equation [2] does not address. Multiplicity distributions from a quantum field theory can be derived using the Alterelli-Parisi equation [22] (also known as the DGLAP equation)

$$D'(x,t) = \int \mathrm{d}z\, D(z,t)\, P(x/z)\, z^{-1} \tag{1}$$

where the integral runs from x to 1 and the derivative on the left is with respect to t. $D(x,t)$ is the 1-particle inclusive distribution and $P(x)$ is the splitting function that dictates the outgoing distribution of incoming momentum at any given vertex. The variables (z, x) are the momentum fraction of the state in question whereas t is the evolution parameter

‡ Figure is borrowed from reference [18].

measuring the virtuality of the initial state. 1-particle inclusive distribution $D(x)$ describes the probability of producing 1 particle with momentum fraction x.

The physical idea is that we start out with a state that has very high virtuality and, as it branches into more and more particles, the final state approaches the mass shell (Fig. 1). There is a renewed interest in the DGLAP equation in recent years as it is used to study inclusive production and branching in conformal field theories [23].

An interesting note here is that the evolution equation for the multiplicity distribution can also be derived as a stochastic branching process. The evolution parameter t can in this sense be treated as a "time" parameter. As discussed in [1] and [18], the φ^3 branching process simply follows the Furry branching process of pure birth. The Furry distribution has been used rather successfully to model particle production through geometric models [19,20]. In fact, it is one of the limiting cases of Giovannini's branching model. However, it still has limitations described previously pertaining to limited physical applicability at higher energies. It nevertheless proves the strength of stochastic methods to modelling particle production. It is this approach that we shall follow in our current work. A more detailed analysis using the DGLAP equation is left to a future study.

3 The Evolution Equation

There are three fundamental processes possible in our model dictated by the same fundamental vertex: ψ Bremsstrahlung, ψ* Bremsstrahlung and ψ* ψ pair production.

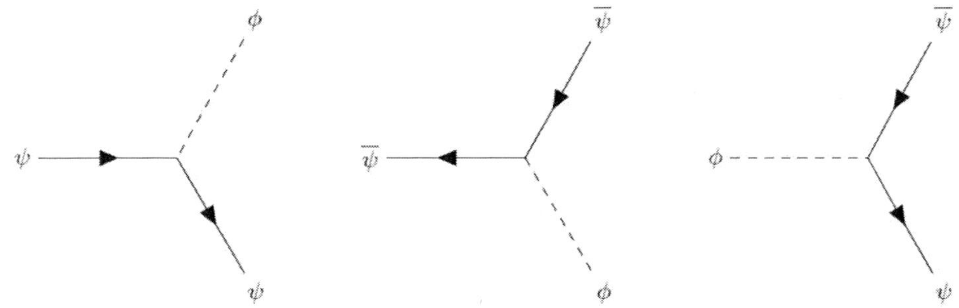

Fig. 2. The three branching processes allowed in our model.

The equation for stochastic evolution of this system can be easily derived. We assign a probability to each of the three kinds of branching processes: A for ψ Bremsstrahlung, A^* for ψ* Bremsstrahlung and B for ψ* ψ pair production. A state with n particles, m antiparticles and o scalars has probability given by P_{mno} and evolves in a small step dt as

$$\begin{aligned} P_{mno}(t + \mathrm{d}t) = &(1 - nA\,\mathrm{d}t - mA^*\,\mathrm{d}t - oB\,\mathrm{d}t)\, P_{mno}(t) \\ &+ nA\,\mathrm{d}t\, P_{mn,o-1}(t) + mA^*\,\mathrm{d}t\, P_{mn,o-1}(t) \\ &+ (o + 1)\, B\,\mathrm{d}t\, P_{m-1,n-1,o+1}(t)\,. \end{aligned} \tag{2}$$

Therefore,

$$\begin{aligned} P'_{mno}(t) = &\, nA\,(P_{mn,o-1} - P_{mno}) + mA^*\,(P_{mn,o-1} - P_{mno}) \\ &+ B\,((o + 1)\, P_{m-1,n-1,o+1} - o\, P_{mno}), \end{aligned} \tag{3}$$

where we have suppressed the t dependence on the right-hand side for brevity. This equation does not have a closed form analytic solution. We therefore use numerical methods to solve it.

There is a small caveat to this distribution. In a high energy experiment, what we measure in the end state is the number of charged particles. Since we wish to model this charged particle multiplicity distribution P_n, we need to sum over all neutral scalars and count all charged particles. P_n is given by

$$P_n = \sum_{i+j=n} \sum_o P_{ijo} \,. \quad (4)$$

Another important note here is that for particle-particle and particle-antiparticle interactions, our initial condition always contains 2 initial charged particles. Our model therefore produces non-zero values only for even multiplicities since all branching processes increase the particle number by 2. This is expected as these numbers are for full phase space, i.e., $-\infty < \eta < \infty$[§]. Since most pp and $\bar{p}p$ data is restricted phase space due to detector design, we introduce a "loss" function to account for particles lost down the beam line.

4 The Loss Function

To account of particles lost outside the detector region and down the beam line, we can perform a convolution of our multiplicity distribution with a probability distribution that describes the loss. The final multiplicity distribution $\bar{P}_n$ should be calculated as follows:

$$\bar{P}_n = \sum_i P_{n+i}\, p_i \quad (5)$$

where p_i is the probability of losing i particles. Following Occam's razor, we assume the simplest form of p_i:

$$p_i = \text{Binomial}\,(i, n+i, p) \quad (6)$$

where Binomial $(i, n+i, p)$ is the probability mass function of the binomial distribution. p_i, therefore, gives us the probability of losing i particles given that $n+i$ particles are produced assuming that p is the probability of losing the particle. Employing such a loss function gives us non-zero values for all multiplicities, as is observed in restricted phase space data for pp and $\bar{p}p$ collisions. Ideally, what we expect is that this probability of loss will increase as we constrain our pseudorapidity window since more and more particles go undetected.

5 Method for Numerical Integration

As mentioned earlier, the evolution equation does not have any closed form solution. Numerical methods are therefore required to further study our model and make contact with data. Numerical integration was done by performing 4th order Runge-Kutta integration (RK-4) for 4 different values of the step-size (h = 0.1, 0.05, 0.01, 0.005) and then using polynomial interpolation to find the values for h = 0. This was done mainly to improve normalisation as RK-4 regularly under-predicts the solution. This error is dependent on the step size h and therefore, treating the total probability of the final state found by the integrator as a function of h should give better results through interpolation. Indeed, this

[§] The UA5 data used for the fit is for a restricted pseudorapidity window. However, the full phase space data reported by UA5 is all even multiplicities as well, as anticipated by our distribution.

resulted in significant improvements in normalisation while saving on computational complexity.

Most of the numerical work was done in Python. However, C++ code was wrapped using Simplified Wrapper and Interface Generator (SWIG) [24] for performance and efficiency gains whenever required. For example, integrating from $t = 0$ to $t = 4.0$ took ~ 43 minutes in Python. The same calculation with wrapped C++ code in Python took ~ 13 seconds.

6 Results and Discussion

The following figure shows the effect of having different values of A and A^* on the multiplicity distribution for ψ ψ interactions and ψ* ψ interactions.

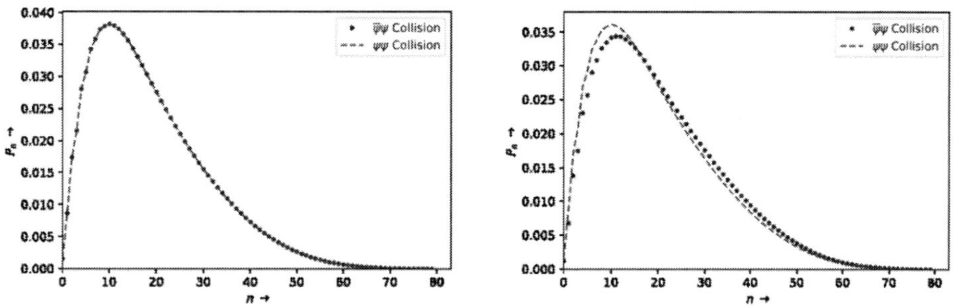

Fig. 3. Multiplicity plots when $A = A^* = 0.8$ and $A = 0.8, A^* = 0.9$.

This is one of the main highlights of this work: from these plots we can conclude that in our model, matter-antimatter asymmetry shows up as a difference in the multiplicity distributions for the particle-particle and particle-antiparticle collisions. Some preliminary fitting using $\bar{p}p$ data from the UA5 collaboration [25] for pseudorapidity window $-3 < \eta < 3$ gives us very different values for A and A^* as is tabulated below. For this fit, we get a very promising value for χ^2 / d.o.f. of 53.27/75. Our model, therefore, successfully captures the feature we were probing for. A detailed analysis of this result should include a fit to pp data from the LHC for the same pseudorapidity window and same energy. This is one of the approaches currently under consideration.

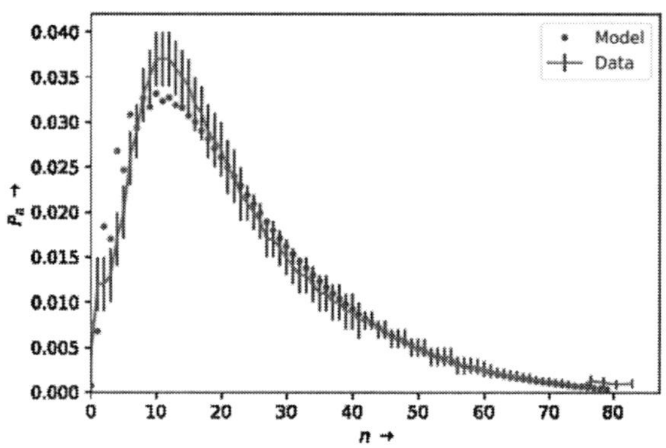

Fig. 4. Preliminary fit of our model to data from UA5 at $\sqrt{s} = 900$ GeV.

Table 1. Model parameter values.

A	A^*	B	p	t
0.19	0.774	0.74	0.186	5.6

7 Conclusion

A phenomenological branching model for particle production that captures matter-antimatter asymmetry is presented. The stochastic equation is solved numerically and preliminary fits to data from the UA5 collaboration seem promising. Future work on the model would involve performing Monte Carlo Markov Chain fitting to do Bayesian parameter estimation and fitting to other datasets. It is expected that the parameter p in the "loss" function should depend only on the pseudorapidity window over which the multiplicity is measured. Further, fitting to pp data from the LHC for the same energy and pseudorapidity window should prove to be a good test for the model. Once the model is rigorously tested against datasets, there are all kinds of phenomenological studies that can be performed. An interesting direction of recent study has been towards modified combinants of multiplicity distribution [26-28]. Some preliminary calculations show promising results from our distribution towards these interesting quantities. These directions are currently being explored.

References

1. R. C. Hwa, *Hadronic Multiparticle Production, ed. P. Carruthers* p. 556 (World Scientific, 1988)
2. A. Giovannini, Nucl. Phys., **Sec B 161** p. 429 (1979)
3. M. Biyajima, Prog. Theo. Phys. **69** p. 966 (1983)
4. M. Biyajima, Phys. Lett. **B137** p. 225 (1984)
5. M. Biyajima, Phys. Lett. **B139** p. 93 (1984)
6. M. Biyajima and N. Suzuki, Phys. Lett. **B143** p. 463 (1984)
7. M. Biyajima, K. Shirame and N. Suzuki, Phys. Rev. **D37** 1824 (1988)
8. N. Suzuki, M. Biyajima and G. Wilk, Phys. Lett. **B268** p. 447 (1991)
9. B. Durand and I. Sarcevic, Phys. Lett. **B172** p. 104 (1986)
10. B. Durand and I. Sarcevic, Phys. Rev. **D36** 2693 (1987)
11. C.C. Shih, Phys Rev. **D33** 3391 (1986)
12. P. Carruthers and C.C. Shih, Int. J. Mod. Phys. **A2** 1447 (1987)
13. S. Sakai, Phys. Rev. **D40** 1430 (1989)
14. V. Gupta and N. Sarma, Z. Phys. **C52** p. 53 (1991)
15. R. C. Hwa, Preprint. Oregon, OITS-374 (1987).
16. R. C. Hwa, Preprint. Z. Phys. **C38** p. 277 (1988)
17. A.H. Chan and C.K. Chew, J. Math. Phys. **33** 3209 (1992)
18. R. C. Hwa, *Intermittency in the φ^3 Branching Model* Nucl. Phys., **Sec B 328** p. 59 (1989)

19. W. R. Chen and R. C. Hwa, Phys. Rev. **D36** p. 760 (1987)
20. W. R. Chen and R. C. Hwa, Phys. Rev. **D39** p. 179 (1989)
21. T. T. Wu, *Scattering and Production at High Energies* Proceedings of the Conference in Honor Of C N Yang's 85th Birthday p. 112 (2008)
22. G. Alterelli and G. Parisi, *Asymptotic Freedom in Parton Language* Nucl. Phys., **B 126** p. 298 (1977)
23. R. Nally, T. Raben, C.-I Tan, *Inclusive Production through AdS/CFT* JHEP **2017**, 75 (2017)
24. SWIG: http://www.swig.org/
25. The UA5 Collaboration, *Proceedings for 24th International Conference on High-energy Physics* p. 647 (1988)
26. G. Wilk and Z. Wodarczyk, *How to retrieve additional information from the multiplicity distributions* J. Phys. G: Nucl. Part. Phys. **44** 015002 (2017)
27. H. W. Ang, A. H. Chan, M. Ghaffar, M. Rybczyński, G. Wilk, Z. Włodarczyk, *Modified Combinant Analysis of the* e^+e^- *Multiplicity Distributions* Mod. Phys. Lett. **A 34** No. 39 (2019)
28. H. W. Ang, A. H. Chan, M. Ghaffar, M. Rybczyński, G. Wilk, Z. Włodarczyk, *A Look at Multiparticle Production via Modified Combinants* EPJ **A 56** 117 (2020)

Effects of phase space variables on oscillations of modified combinants

H.W. Ang*, *Z.* Ong, *P.* Agarwal, *A.H.* Chan, and *C.H.* Oh

[1]Department of Physics, National University of Singapore

Abstract. It has been shown recently that additional information can be obtained from charged particle multiplicity distribution by investigating their modified combinants C_j, which exhibit periodic oscillatory behaviour. The modified combinants obtained from experimental data can be expressed in a recurrent form involving the probability of obtaining N charged particles $P(N)$, scaled by the void probability $P(0)$. The effects of various experimental observables such as $|\eta|$, p_T and centre-of-mass collision energy $\sqrt{s}$ on the oscillations of C_j will be discussed.

1 Introduction

The "inner-space outer space connection" was one of the postulates conceived in the early 1990s to explain the numerous connections between the fields of particle physics and cosmology. One of these connections manifests in the form of multiplicity distribution (MD) of charged hadrons to the distribution of observed galaxies, both of which are products of the process of hadronization. Another similarity lies in the study of void probability $P(0)$ [1], which represents the probability of observing no galaxies within a certain region of space. In the context of multiparticle production, $P(0)$ represents the probability of observing no hadrons within a certain region of phase space.

In this work, we introduce the notion of modified combinant C_j to be used in the analysis of MD. The C_j's derived from experimental data have been observed to undergo periodic oscillations as a function of rank j. Their relationship to $P(N)$, as well as the effects of pseudorapidity η, transverse momentum p_T and centre-of-mass collision energy $\sqrt{s}$ on their oscillation will be explored.

2 Relationship between C_j and $P(N)$

Apart from the generating function and probability function, information in a statistical distribution can also be expressed in a recurrent form involving only adjacent values of probability $P(N)$

$$(N+1)P(N+1) = g(N)P(N). \tag{1}$$

*e-mail: ang.h.w@u.nus.edu

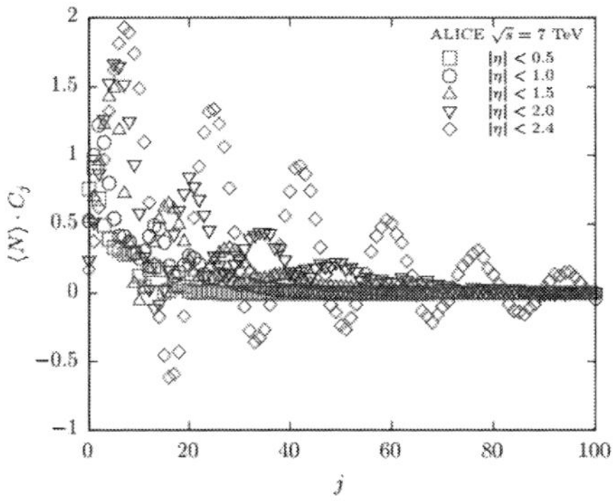

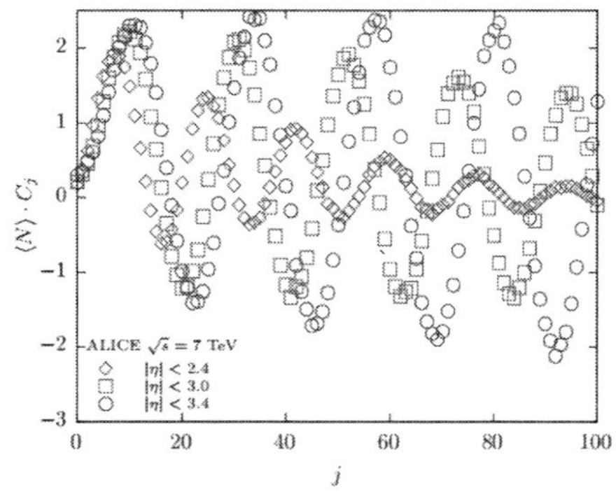

Figure 1. Plot of C_j derived from ALICE data. **Left panel:** The plot is made with MD data up to $|\eta| < 2.4$. The amplitude increases with wider pseudorapidity interval which decays with increasing j. **Right panel:** The plot made with data up to $|\eta| < 3.4$ shows that the amplitude decay stagnates at $|\eta| < 3.0$ and increases with increasing j at $|\eta| < 3.4$.

A linear form of $g(N)$, though useful, is somewhat limited in the sense that any $P(N)$ is only related to the value of $P(N-1)$. This constraint seems unnecessarily restrictive. In [2], it is proposed that Eqn (1) be modified in order to factor effects of all $P(j)$'s for $j < N$

$$\langle N \rangle C_j = (j+1)\left[\frac{P(j+1)}{P(0)}\right] - \langle N \rangle \sum_{i=0}^{j-1} C_i \left[\frac{P(j-i)}{P(0)}\right] \tag{2}$$

The coefficients C_j in Eqn (2) are the modified combinants. In this form, it is clear that C_j acts as the weight that determines the relative contributions of all smaller $P(N-j)$'s to the value of $P(N)$. In some sense, one can interpret C_j as the "memory" that the $P(N)$ term has of all lower multiplicity terms.

3 Dependence of C_j oscillations on phase space variables

In high-energy collision experiments, the phase space is typically characterised by three variables - η, p_T and $\sqrt{s}$. Each of these variables contain information of the characteristics of the collision, which affect the probability of producing a certain number of particles. On this note, we will study the effects of each of these three variables on the oscillation of C_j.

3.1 Dependence on η

The pseudorapidity variable, denoted by η, characterises the angle relative to the beam axis. The prefix "pseudo" stems from the fact that it is used as a proxy for rapidity (typically denoted by y). The experimental calculation of rapidity involved knowing the longitudinal component of a particle's momentum p_z and its energy E, quantities which are not easily available. As such, the pseudorapidity variable η defined in place as a function of θ, the angle between the forward beam direction and the particle's 3-momentum. This is a more experimentally accessible quantity and in high-energy physics, where the momentum is much larger than the rest mass of a particle, $\eta \approx y$.

The behaviour of C_j plotted as a function of j from experimental data published by the ALICE collaboration [3, 4] is shown in Fig 1. It is evident that both the period and amplitude

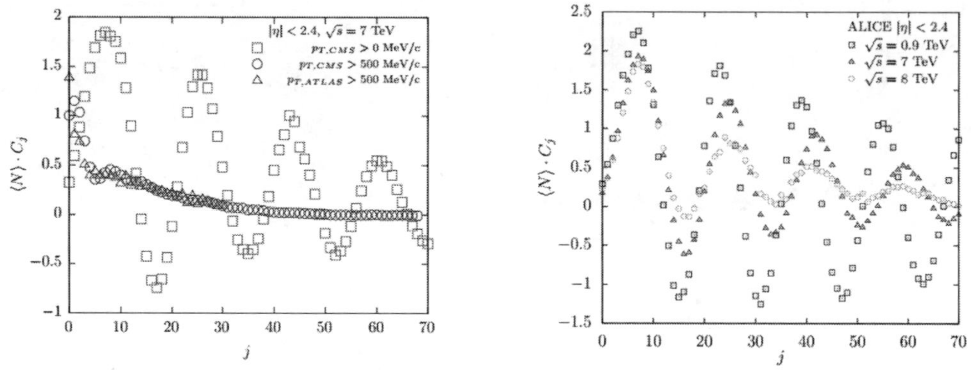

Figure 2. Left panel: The C_j oscillations across different transverse momentum cuts from data in [5]. Once again, data from a larger phase space volume of $p_T > 0$ MeV/c exhibits more significant oscillation. **Right panel:** C_j plotted across various centre-of-mass energies at the same pseudorapidity interval of $|\eta| < 2.4$ from ALICE data.

of the oscillations increase with the pseudorapidity interval. Comparing the left and right panels of Fig. 1, one observes that the rate of amplitude decay decreases at larger pseudorapidity interval. In fact, the amplitude seems to remain constant at $|\eta| < 3.0$ before reversing and increases for $|\eta| < 3.4$.

3.2 Dependence on p_T

In high-energy collisions, transverse momentum, or p_T, is the component of the particle's 3-momentum that is perpendicular to the beam axis. It provides an indication of the "hardness" of the underlying interaction in the collisions.

In general, harder collisions give rise to particles with higher p_T while particles resulting from softer collisions possess lower p_T. Hard (or short-distance) interactions can be accurately described by perturbative QCD (pQCD) in increasing powers of the coupling constant $\alpha_s < 1$ in this regime. Unfortunately, such perturbative techniques fail in describing the physics of soft (or large distance) interactions, which corresponds to $\alpha_s > 1$.

A reference to a plot of p_T vs N (such as that in [5]) will show that soft interactions produce majority of the detected secondary particles as compared to the hard interactions. This obliges the introduction of various phenomenological models to augment pQCD in providing a complete picture of the underlying physics.

Such a characteristic makes the study of the dependence of C_j oscillations on p_T of interest. In left panel of Fig. 2, the pseudorapidity interval[1] and centre-of-mass energy variables are kept constant to isolate the effects of changing p_T on C_j. Oscillations are only observed for the experimental data extrapolated to $p_T > 0$ MeV/c from the CMS collaboration, which includes more secondary particles from the soft collisions.

3.3 Dependence on $\sqrt{s}$

Centre-of-mass energy, denoted by $\sqrt{s}$, provides an indication of the maximum amount of energy available for exchange in interactions during a high-energy collision. In a nutshell,

[1]The data from the ATLAS collaboration is taken over a slightly wider interval of $|\eta| < 2.5$ compared to $|\eta| < 2.4$ from CMS. However, as observed in the left panel of Fig. 2, the slightly wider interval makes no practical difference at $p_T > 500$ MeV/c.

collisions at a higher $\sqrt{s}$ has a higher probability of producing more particles due to the mass-energy relation $E^2 = p^2 + m^2$. In this way, the multiplicity of the collisions is also a function of $\sqrt{s}$.

Once again, in the right panel of Fig. 2, the pseudorapidity interval has been kept constant at $|\eta| < 2.4$ to isolate the effects of a changing $\sqrt{s}$. It is observed that the lower the $\sqrt{s}$, the larger the amplitude of C_j oscillations, and the slower the rate of amplitude decay. However, there is no clear trend in the period of the oscillation with increasing energy.

4 Discussion of results

It is clear that each of the three experimental variables discussed has effects on the oscillations of C_j, and by extension the fluctuation of "memory" on lower multiplicities. In Sect. 3.1, we see that the oscillations change violently with η. In general, the larger the $|\eta|$ interval, the larger the oscillations and the slower the oscillations decay. This seem to indicate that a larger region of available phase space corresponds to more violent oscillations in C_j. In a way, one can view information captured from a larger $|\eta|$ interval as being more indicative of the total available information of the underlying physics. This can simply be explained from the very fact that a wider $|\eta|$ interval captures more of the produced particles, as well as decays from leading particle.

The tentative hypothesis on larger available phase space leading to more violent oscillations is also supported by the plot in the left panel of Fig. 2. In this plot, the cut made at $p_T > 0$ MeV/c will undoubtedly capture more of the underlying produced particles than the cut made at $p_T > 500$ MeV/c, with a greater available phase space volume in the former. Once again, this leads to a more violent oscillation in C_j with the increased phase space volume.

Interestingly, the dependence of oscillation on $\sqrt{s}$, as depicted on the right panel of Fig. 2, does not corroborate with the hypothesis proposed above. If one were to restrict the definition of phase space variables to that involving the spatial (proxied by $|\eta|$) and momentum (p_T) variables, then the $\sqrt{s}$-dependence should not be interpreted in the light of the hypothesis, which describes only dependence on phase-space variables.

This study elucidates the dependence of C_j oscillations on availability of phase space, or extent of the available phase space volume. In general, the larger the available phase space, the more violent the C_j oscillations. A tentative phenomenological model has been proposed to explain the existence of such oscillations in [6]. Model description of the fluctuating C_j is possible if one includes the binomial distribution as one of the two distributions to be compounded in a compound distribution.

Comparison of the differences in the period of oscillation and increase in amplitude also leads to a proposal of using C_j oscillations to distinguish between matter-matter and matter-antimatter collisions as suggested in [7].

5 Conclusion

The dependence of C_j oscillations on η, p_T and $\sqrt{s}$ are partially explored in this work. It is seen that C_j provides an alternate means to interpret the multiplicity data obtained from high-energy collision experiments under various conditions. As mentioned in section 1, if the inner-space-outer-space connection holds true, C_j may also be considered a potential candidate as a tool in the study of galaxy distribution in our universe. In the vast expanse of space, quantities such as the void probability $P(0)$ should be obtainable via observation and with it, calculations on C_j based on observational data can be performed.

The utility of C_j as introduced in this work is far from complete. In [7], links between C_j and various moments, as well as to the non-extensive parameter q in Tsallis statistics [8], have been made. Hence, there is much to be explored in the full range of utility of C_j, this work of which is only the tip of the iceberg.

Acknowledgement

We (H.W. Ang, Z. Ong and P. Agarwal) will like to thank the NUS Research Scholarship for partially funding this study. One of us (H.W. Ang) will like to thank Prof. G. Wilk, Prof Z. Włodarczyk and Prof M. Rybczyński for their critical discussions and comments.

References

[1] J. N. Fry, Astrophys. J **289**, 10-17 (1985)
[2] B. E. A. Saleh and M. C. Teich, IEEE Proc **70**, 229-245 (1982)
[3] S. Acharya *et al.*, EPJC **77**, 852 (2017)
[4] J. Adam *et al.*, EPJC **77**, 33 (2017)
[5] V. Khachatryan *et al.*, JHEP **01**, 79 (2011)
[6] H. W. Ang *et al.*, MPLA **34**, 1950324 (2019)
[7] H. W. Ang *et al.*, arXiv preprint, arXiv:1908.11062 (2019)
[8] C. Tsallis, J. Stat. Phys **52**, 479-487 (1988)

Investigation of the Sudden Solar Ionospheric Disturbance using Radio Telescope

Adam Aqasha[1], *Andrien* Zheng[1], *Sneha* Athreya[2*], and *Hoe Teck* Tan[1]

[1]School of Science and Technology, 1 Technology Drive, Singapore
[2]National Junior College, 37 Hillcrest Road, Singapore

Abstract. Low-frequency radio telescopes are cheap and useful devices for the investigation of terrestrial and extra-terrestrial emissions. These emissions come either from the Sun and the planet Jupiter to terrestrial emissions. This project aims to investigate the Very Low Frequency (VLF) waves from mid-August to October 2019 using Radio JOVE (20 MHz) and SSID (3-30 kHz) to observe for the occurrence of solar flares and see how if the radio telescopes that the team set up is reliable. This will allow us future students aspiring to learn about astronomy to examine solar flares in detail during the upcoming solar maximum. Not many flares were detected as this period happens to be a solar minimum. However, a series of flares occurred between 30 September 2019 and 1 October 2019, which the telescopes have been able to detect, particularly SSID.

1. Introduction

Solar events have great influences over the Solar System. On Earth, the understanding of these events is paramount. They produce streams of high energy particles which greatly affect the Earth's space weather and may cause hazards to the space-stations, the spacecrafts and the astronauts. In addition, massive solar events may also affect the terrestrial systems. They sometimes post dangers to the satellite systems as well as causing blackouts to the electric power grids.

In this paper, we focus on the solar activities like the solar flares and the coronal mass ejections. These activities cause disturbances to the ionosphere. The ionosphere is responsible for the reflection of radio waves. Thus, any disturbance to the ionosphere will affect the reflection of the radio waves. We investigate how these solar events affect the radio waves in the Very Low Frequency regime.

1.1 The Ionosphere

The ionosphere is an ionized layer in the atmosphere that is around 50 to 600km above the Earth's surface. It is caused by the UV radiation that comes from the Sun. The degree of ionization changes due to many factors, such as latitude, height above ground or time of the day. During the day, UV radiation that is received from the sun is high, causing a greater degree of ionization, while at night, the electrons and ions recombine and the level of ionization decreases. There is also a greater degree of ionization closer to the equator as these

* Email: snehaathreya03@gmail.com

regions receive a higher and more concentrated amount of UV radiation from the Sun. The ionosphere consists of 3 main layers-D, E and F. The following table describes the properties of these layers:

Table 1: Layers of the Ionosphere

Layer	Altitude	Time it is present	Primary ionizing radiation	Effect on VLF
D	60-90km	Day	Hard X-rays (0.1-1nm)	Absorption/ Reflection
E	100-125km	Day and night	Soft X-rays (1-20nm), or Extreme UV (80-103nm)	Reflection
F	300-400km	Day and night	Extreme UV (20-80nm)	Reflection

The ionosphere is a great propagator of radio waves. Monitoring the power of these radio signals allows us to detect changes in the ionosphere. When a solar storm occurs, the D region of the ionosphere drastically reaches a high level of ionization, allowing greater reflection of the Very Low Frequency (VLF) waves. This results in the radio signals showing a sudden, sharp spike upwards. We monitor VLF as it is readily available all over the world and because it is transmitted at constant power. [1, 5, 6]

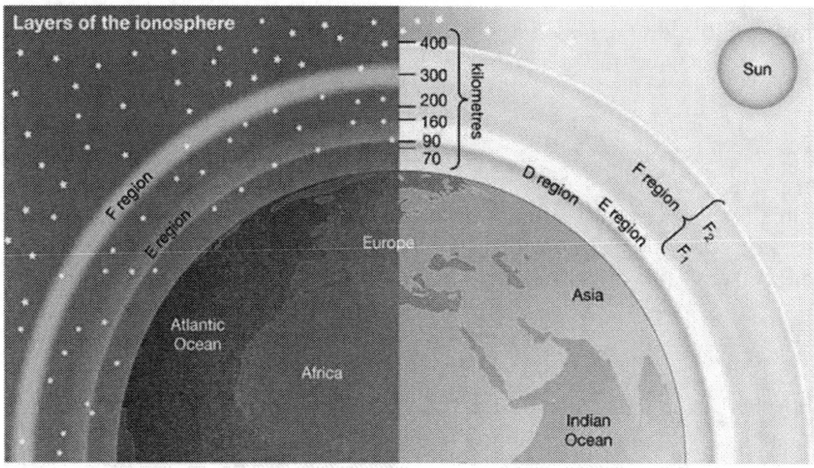

Figure 1: Layers of the Ionosphere [11]

1.2 Solar Flares

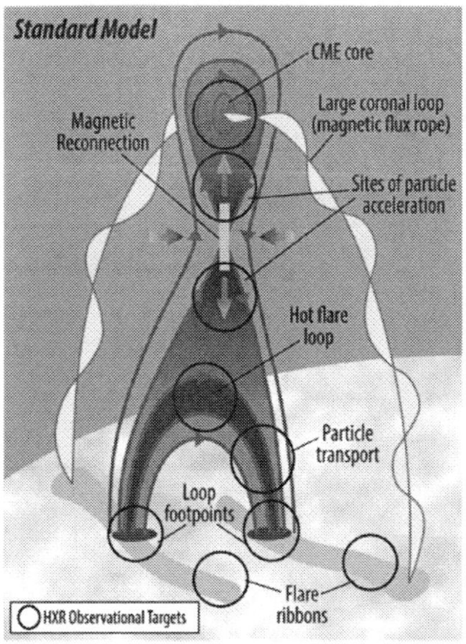

Figure 2: Model of Solar Flare [12]

Solar flares are massive explosions on the Sun that occur when charged particles in the Sun's plasma erupt into space, travelling with enormous speed. These flares occur when the magnetic field lines of the sun reconnect, as seen in Figure 2, releasing large amounts of energy and X-ray waves into space. It usually occurs in the active regions of the sun, near sunspots, along the dividing line between areas of oppositely directed magnetic fields, where temperatures might even reach more than 10,000 Kelvin. Depending on the location and trajectory of the solar flares, the X-rays that are shot out may create electrical disturbances in the ionosphere of the Earth. The process of the X-rays colliding with earth and disturbing its ionized layer is known as solar storms.

One of the huge dangers of a solar flare is widespread electrical disruption. When the particles strike the Earth's magnetosphere, they could produce an electrical charge, an electrical charge that is powerful enough to travel to the planet's surface. When those charged currents encounter electrical grids, it may lead to complications such as power faults which may affect cities or even whole countries. On 12 March 1989, a particularly strong solar flare struck North America and overwhelmed the electrical grid of the Canadian province of Quebec. At 2:44 a.m. the following morning, a series of cascade failures occurred in the electrical system, resulting in a province-wide blackout that lasted 12 hours.

We believe that a future solar storm might be even more devastating as there are a lot greater advancements in technologies currently as compared to 1989. This makes us more vulnerable as more systems depend on electricity to function properly. Due to the immense impact these solar storms can have on business, telecommunication systems, and other aspects of our life, we found that ionospheric disturbances are an interesting study and hence the focus of our project [2, 5, 6].

The different categories of solar flares are:

Table 2: Classes of Solar Flares

Solar Flare	X-ray flux (Log Watts/ Sq. m)	Effect
X	-4	Causes very serious damage to radio communications, even blackouts.
M	-5	Causes some impact on radio communications.
C	-6	Minor
B	-7	None
A	-8	None

1.3 Coronal Mass Ejections (CME)

During Coronal Mass Ejections (CMEs), enormous bubbles of superheated gas – called plasma – are ejected from the Sun. Over several hours, they can eject billions of tons of coronal material off the Sun's surface and accelerated to speeds of a million miles per hour (1.6 million kilometres per hour) and carry an embedded magnetic field (frozen in flux) that is stronger than the background solar wind interplanetary magnetic field (IMF) strength. The fastest Earth-directed CMEs can travel to our planet in as short as 15-18 hours. Slower CMEs might take several days to arrive. They expand in size as they propagate away from the Sun and larger CMEs can reach a size comprising nearly one-quarter of the space between Earth and the Sun by the time it reaches our planet. CMEs can happen several times a day when the Sun is most active, whereas, during its fewer active periods, CMEs could occur only about once every five days [3, 6].

2. Aim

Our aim for the project is to test if the telescopes installed accurately receive signals and can detect the occurrence of solar flares.

3. Methodology

A Radio JOVE (20 MHz) and SSID (3-30 kHz) receiver was connected to the computer directly to continuously monitor the ionosphere. Radio JOVE was set up (figure 1) at National Junior College (NJC) while the SSID was placed at the School of Science and Technology (SST). The signal strength was monitored for any disturbances and counter checked with data from earth and space weather forecasts. The archived data would then be used to confirm our results about the disturbances.

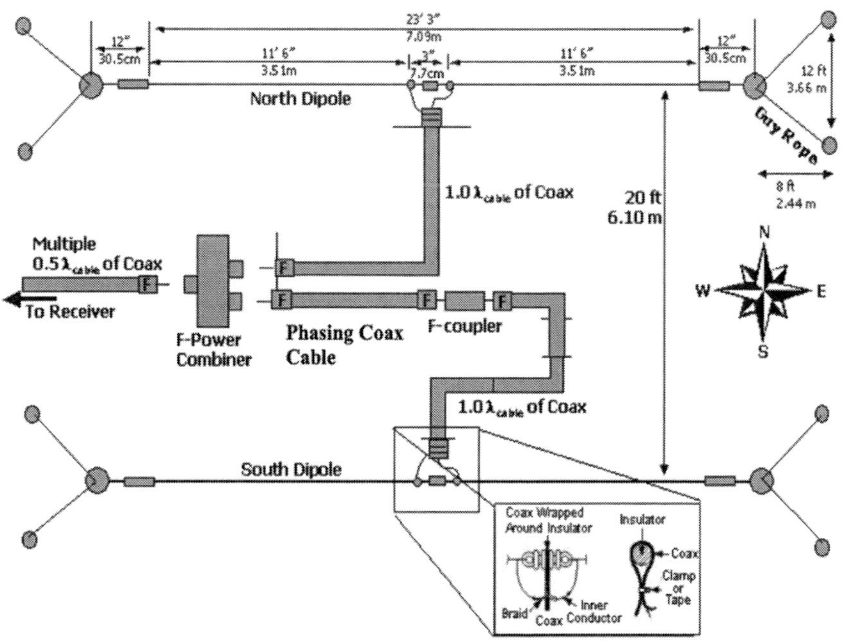

Figure 1: Setup of Radiojove
(source: https://radiojove.gsfc.nasa.gov/telescope/ant_manual.pdf) [4]

Figure 2: Deployment of Radiojove

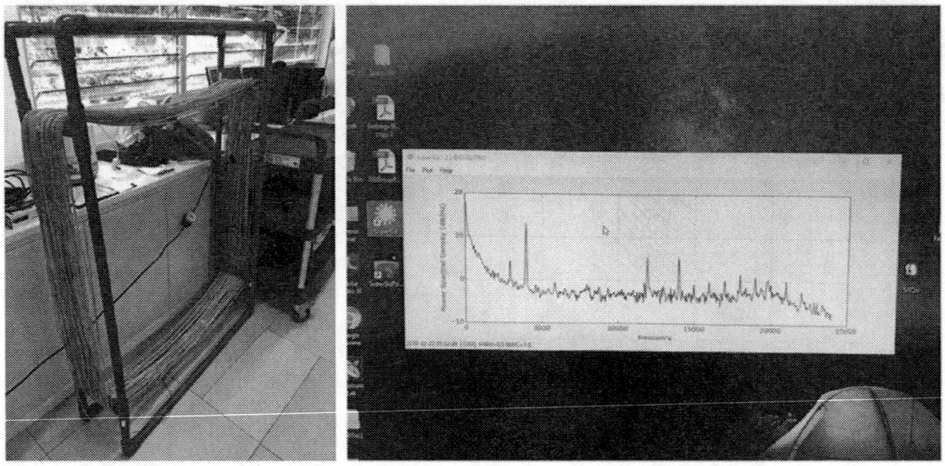

Figure 3: Deployment of the SSID

4. Results and Discussion

The following are the graphical images obtained using the Skypipe 2 software for Radio JOVE and software for SSID. Since the period of data collection happens to be during a solar minimum, there was a lack of flares observed. However, a small series of flares may have been detected by the telescopes from 30 September 2019 to 1 October 2019, of which the results shall be shown.

The graphs for Radio JOVE show changes in radio signal against local time in Singapore, while SSID shows the change in radio signals against Coordinated Universal Time (UTC), which is 8 hours behind Singapore.

Table 3: Graphs showing strength of radio signals recorded by Radio JOVE and SSID

<table>
<tr><td>Date</td><td>Radio JOVE</td><td>SSID</td></tr>
<tr><td>29 Sep</td><td>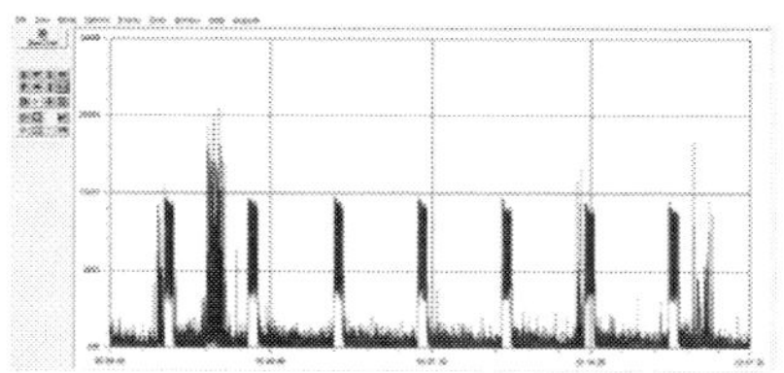</td><td>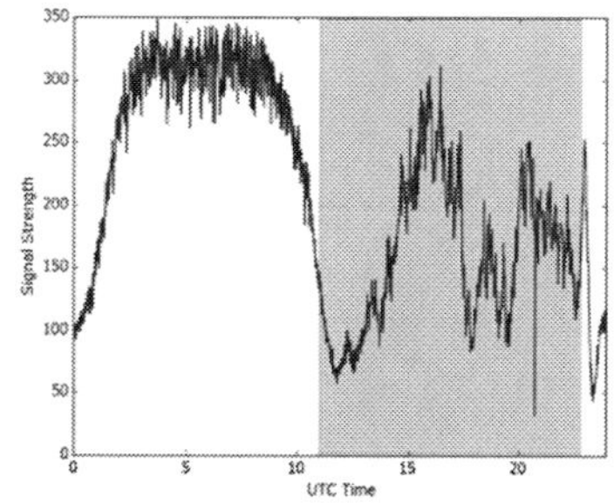
</td></tr>
<tr><td colspan="3">Maximum flux: A8.64
No flares reported by https://www.spaceweatherlive.com/
Background flux: A4.7 [7]

Low signal strength recorded by SSID of no more than 350. Some disturbance recorded by Radio JOVE around 8am (0:00 UTC on 29 September).</td></tr>
<tr><td>30 Sep</td><td>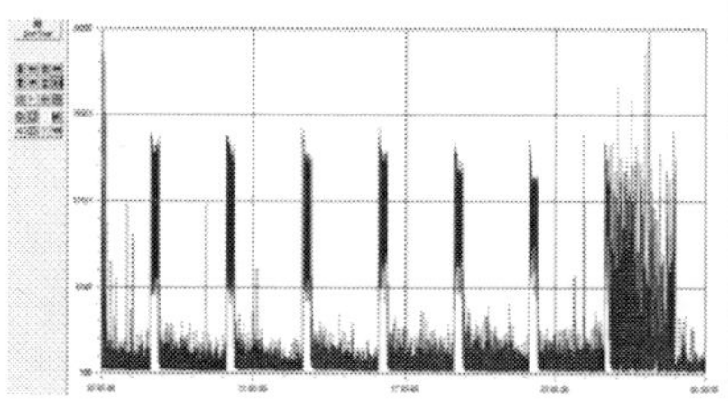</td><td>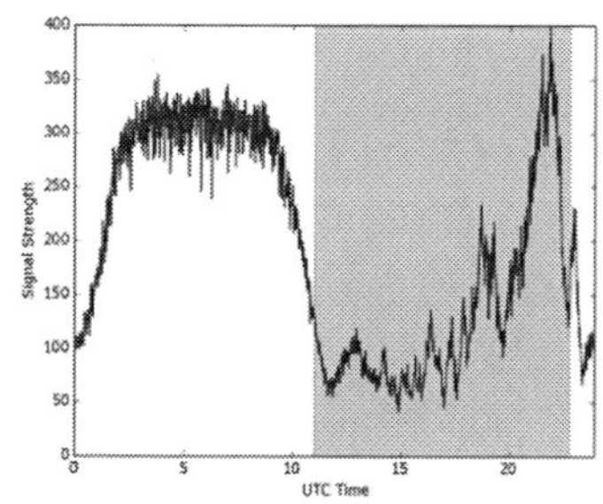
</td></tr>
<tr><td colspan="3">Maximum flux: B1.57
No flares reported by https://www.spaceweatherlive.com/
Background flux: A2.6 [8]

The SSID graph shows a high peak (Signal strength: 400) after 20:00 UTC during the occurrence of the solar flares. 20:00 UTC corresponds to 4am in Singapore on 1 of October, which is 8 hours ahead.

The graph on the left shows results from 5am on 30 September to 5 am on 1 October. There was a high amount of disturbance from around 2am to 4am that could be attributed to the occurrence of a solar flare.</td></tr>
</table>

<table>
<tr><td>1
October</td><td>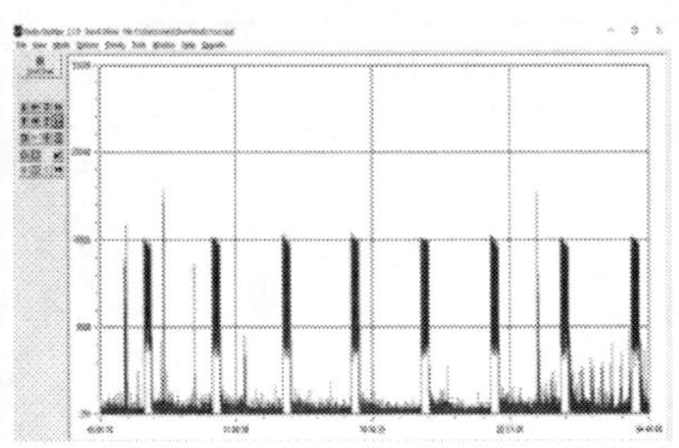</td><td>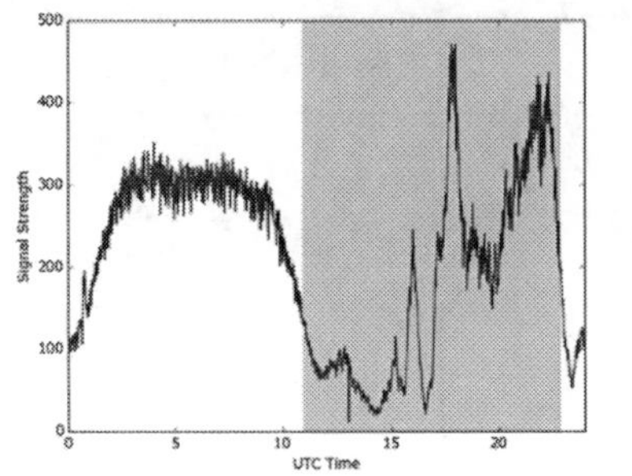
</td></tr>
<tr><td colspan="3">Solar flare no.: 1
Intensity: B1.67
Start time:1.41 UTC
Max: 1.45 UTC
End:1.49 UTC
Background flux: A2.3 [9]

Our results for Radio JOVE (left) display the power of the signal from 5am of 1 October to 5 am 2 October on this diagram. Slight disturbance was detected before 12 pm and after 11pm.

The peak point for the SSID graph (right) is observed at around 18:00 UTC, where signal strength reached an unnaturally high level of 470.</td></tr>
</table>

Another source that had confirmed the occurrence of the solar flares at this time was ISRO's (Indian Space Research Organization) website, a more reliable source than https://www.spaceweatherlive.com/ [10].

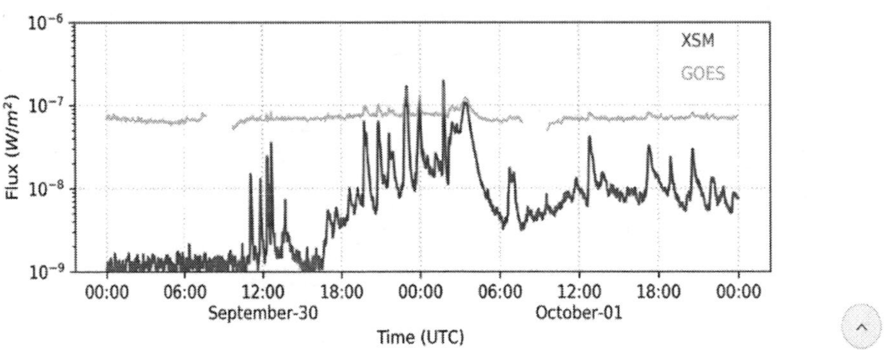

Figure 4: Graph showing the solar X-ray flux as measured by XSM (in blue) and flux measured by X-ray sensor on Geostationary Operational Environmental Satellite (GOES-15) (in orange)

Figure 4 shows that there indeed was a series of flares detected from 30th September 2019 00:00 UTC - 1st October 2019 23:59 UTC by the Solar X-ray Monitor (XSM) that was carried by Chandrayaan-2 orbiter. Space Weather Live had only identified the occurrence of 1 solar flare during this period; this may have been because the receiver was facing away from the sun while the flare occurred. Singapore was also facing away from the sun when the solar flares began on 29 September (around 12:00 UTC) but the telescopes managed to observe some spikes on 1 October during the daytime. This shows that the telescopes were successful in detecting the flares.

5. Conclusion

Solar flares of B-class were detected by our telescopes. The regular disturbances for the Radio JOVE make it difficult for us to pinpoint the time from which the solar flares may have begun. Due to the disturbances and the uncertainty for what caused them, our results for Radio JOVE may be not as reliable as our results from SSID.

6. References

[1] Tracking Solar Flares. (n.d.). Retrieved 2019, from http://solar-center.stanford.edu/SID/activities/ionosphere.html
[2] Kazmeyer, M. (2019, March 2). What Effects Can Solar Flares Have Directly on the Earth? Retrieved 2019, from https://sciencing.com/effects-can-solar-flares-directly-earth-3864.html
[3] What is a coronal mass ejection? (n.d.). Retrieved July 2019, from https://earthsky.org/space/what-are-coronal-mass-ejections
[4] National Aeronautics and Space Administration. (2012). *Radio Jove RJ1.2 Antenna Kit Assembly Manual*. Retrieved from - https://radiojove.gsfc.nasa.gov/telescope/ant_manual.pdf
[5] Arnold, S. (2014). Getting Started in Radio Astronomy. *The Patrick Moore Practical Astronomy Series*. doi: 10.1007/978-1-4614-8157-7
[6] Andrews, P. (2011). *How to Build Your Own Radio Telescope* (3rd ed.).
[7] Solar flares - Sunday, 29 September 2019. (n.d.). Retrieved 2019, from https://www.spaceweatherlive.com/en/archive/2019/09/29/xray
[8] Solar flares - Monday, 30 September 2019. (n.d.). Retrieved 2019, from https://www.spaceweatherlive.com/en/archive/2019/09/30/xray
[9] Solar flares - Tuesday, 1 October 2019. (n.d.). Retrieved 2019, from https://www.spaceweatherlive.com/en/archive/2019/10/01/xray
[10] Solar flare observed by the Solar X-ray Monitor on Chandrayaan-2. (n.d.). Retrieved December 2019, from https://www.isro.gov.in/update/10-oct-2019/solar-flare-observed-solar-x-ray-monitor-chandrayaan-2

6.1 Images

[11] **https://cdn.britannica.com/s:700x500/46/109746-050-9511BBEF/differences-layers-ionosphere-Earth.jpg**
[12] **https://www.researchgate.net/figure/HXR-observations-emission-from-all-keypoints-in-the-standard-model-of-solar-eruptive_fig1_312061210**

Development of a Radio Telescope for Very Low Frequency Observations

Arpit Gupta[1,*], *Seow* Kit Hint[2], and *Cao* Shangyu[1] and *Hoe Teck* Tan[3]

[1]Student, National Junior College, Singapore
[2]Student, School of Science and Technology
[3]Teacher, School of Science and Technology

Abstract. Sudden ionospheric disturbances are transient changes in the ionosphere caused by enhancement in X-ray and EUV fluxes during solar flare events. The Solar Storm Radio Telescope is developed to detect Very Low Frequency (VLF) signals with frequency between 3-30 kHz transmitted from various VLF stations around the Globe. We will also be investigating different methods to reduce the background noise in the data collection. This will help to ensure an accurate hit when there is a sudden ionospheric disturbance.

1 Introduction

A Solar Storm Radio Telescope is a device that utilises radio waves propagation and sudden ionospheric disturbance to monitor solar storms. It is used to observe and study astronomical phenomena unobservable by the human eye.

Solar storms have a very diverse effect on the earth, examples of such include sunspots influencing the weather, solar flares disrupting satellite signal transmissions to earth and the well-known coronal mass ejection (CME) which behaves very much like an electromagnetic pulse (EMP) given its ability to disrupt and wreak havoc on electronics. Although we know about solar storms and how they affect us, we don't know enough about them to take preventive action against them to safeguard human civilisation should a coronal mass ejection hit the earth. It is precisely because we know so little, about solar flares that we need to study them further using a radio telescope to further deepen our understanding about solar flares and the sun, in the hopes of being able to find a solution to protect human civilisation and its technology should we ever be hit by a solar storm of devastating magnitude.

The SSRT we will be developing is designed to be low cost and easy to construct. Our radio telescope will encourage young aspiring engineers and radio enthusiasts to pick up their first project in radio astronomy due to its low cost making it affordable for anyone from students to adults alike as well as its simplicity which will allow anyone to take on this project as long as they have the fundamental technical knowledge on engineering and radio astronomy. This will also contribute to the effort of raising awareness on how important radio astronomy is to science and humanity in general.

* E-mail: arpitg1011@gmail.com

1.1 Hypothesis

By building a radio telescope with an antenna composed of 12 AWG wire and PVC pipes to pick up radio signals at very low frequencies, an amplifier circuit based around a dual 082 JFET Operational Amplifiers and a 741 type op-amp to amplify and reduce interference of the signal received, a pc with a soundcard and an audio analysing package installed on the pc to plot the data on ionospheric activity in real-time on a chart, one can observe the sun's effect on our ionosphere by studying the fluctuations in ionospheric activity.

1.2 Engineering Goals

To provide amateur radio enthusiasts with a low-cost and self-customisable alternative to buying radio telescopes off the shelf as well as to encourage students and adults to take up engineering if they have yet to do so and are interested through this relatively simple project involving fundamental technical knowledge on radio astronomy and electronics.

2 Procedures

Figures and tables, as originals of good quality and well contrasted, are to be in their final form, ready for reproduction, pasted in the appropriate place in the text. Try to ensure that the size of the text in your figures is approximately the same size as the main text (10 point). Try to ensure that lines are no thinner than 0.25 point.

2.1 Antenna:

1. Loop 200m of 12 AWG wire around a PVC pipe structure of preferably 1.20m x 0.95m x 0.95m.

2. Deduce the natural resonance of the antenna by feeding a sine wave of 5 kHz into a separate loop of the same wire. The voltage induced is measured. The frequency of the signal being fed into the single loop is gradually increased. The natural resonance of the antenna is the frequency at which the voltage peaks.

3. The antenna is then tuned by adding polypropylene capacitors in parallel to the end of the antenna, such that its natural resonance is equal to that of the VLF transmitter that will be used for VLF observation. Alternatively, changing the number of loops of wire of the antenna also changes the natural resonance of the antenna.

2.2 Amplifier Circuit

1. Assemble the circuit on a breadboard(excluding dotted sectors) for testing.
2. Test the circuit by feeding a sine wave from a function generator into terminal block J1, while using an oscilloscope to check for any form of signal amplification

3. Once amplification has been proven successful, assemble the circuit on a stripboard/perf board for durability purposes and to make the circuit more compact.

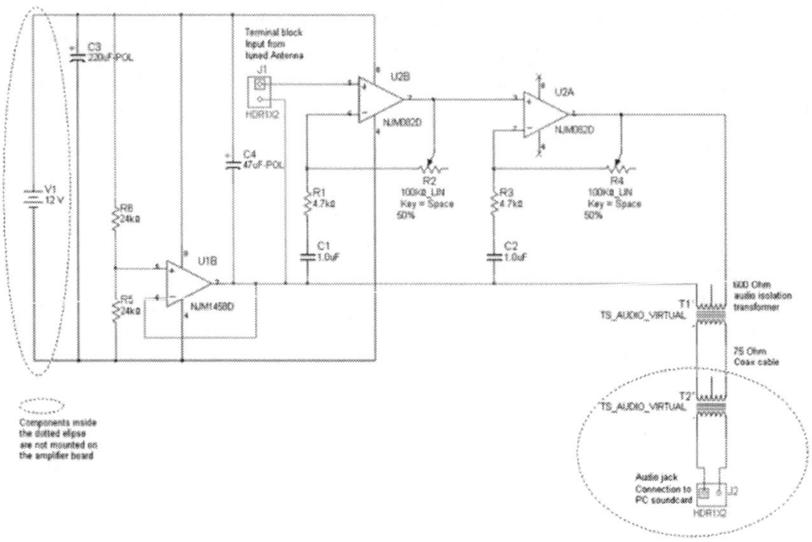

Fig 1: Schematic diagram of amplifier circuit

2.3 Data Collection

'Spectrum Lab' software, a free specialised audio analyser package by 'DL4YHF', is used. It digitises the signal received at the sound card from the antenna and performs a fast Fourier transform on it to produces charts of the power against time by reading the log files and graphing them.

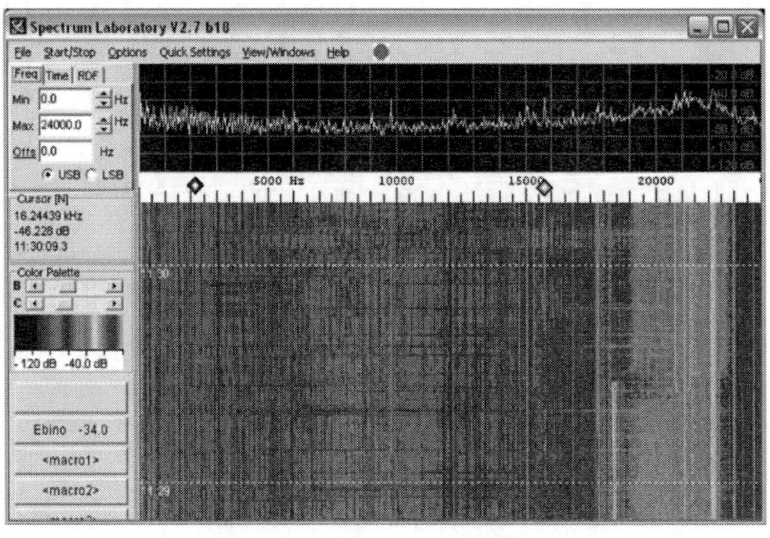

Fig 2: 'Spectrum Lab' software

3 Results

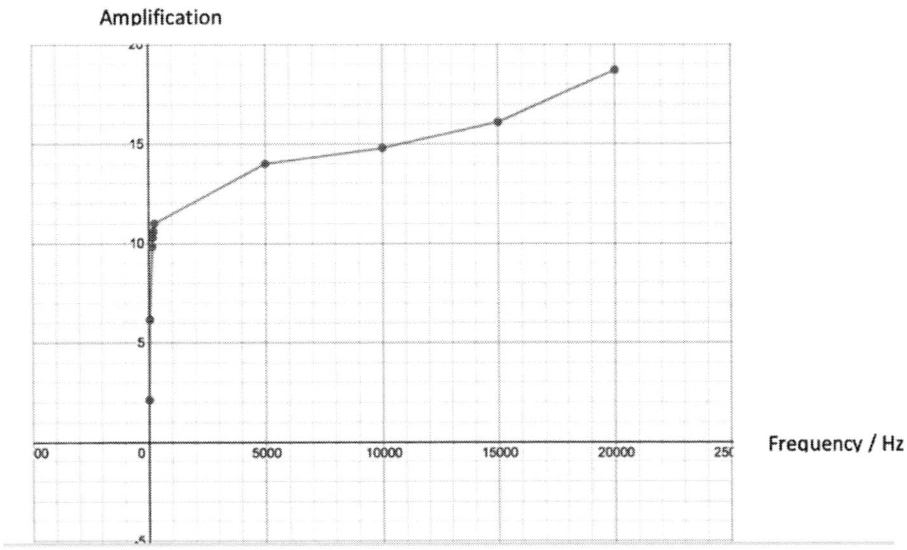

Fig 3: Amplification achieved at various frequencies

3.1 Conclusion

From the results, we can conclude that the construction was successful as the signal's peak to peak was amplified across the wide range of frequencies. However, it can also be observed that as the signal frequency increases, so does the amplification. This could be due to the fact that electromagnetic force (EMF) is proportional to signal frequency.

Circuit construction was successful with breadboard as well as amplifying test turning up positive results. The circuit was able to consistently amplify the signal fed to it at multiple frequencies as such it can be concluded that the circuit assembly was successful.

This implies that such a low-cost and effective do-it-yourself alternative to the rather costly radio telescope can be deemed feasible and may even be adopted by radio enthusiasts and amateur astronomers alike not only for VLF observations but possibly further and better modification as well.

With the low-cost, mass production may be an option and could very well imply the possibility of setting up an SSRT array island-wide should others choose to adopt this idea and cooperate with each other on the sharing of data.

References

1. Percival Andrews, *How to Build Your Own Radio Telescope* (2007)
2. Tracking Solar Flares. (n.d.). Retrieved from, http://solar-center.stanford.edu/SID/activities/ionosphere.html
3. SPACEARTH TECHNOLOGY. (n.d.). Ionospheric physics. Retrieved from, http://www.spacearth.net/?page_id=308

Probing Supercritical Accretion in Ultraluminous X-ray Source M82 X-1 by means of X-ray Spectral Evolution Analysis

Izatul Hafizah[1,*] and *Kiki* Vierdayanti[2,**]

[1]Astronomy Study Program, Faculty of Mathematics and Natural Sciences, Institut Teknologi Bandung, Ganesha 10, Bandung, 40132, Indonesia
[2]Astronomy Research Division, Faculty of Mathematics and Natural Sciences, Institut Teknologi Bandung, Ganesha 10, Bandung, 40132, Indonesia

Abstract. We analyze the spectral evolution of ultraluminous X-ray source (ULX) M82 X-1 by means of spectral fitting. We use selected *Swift*/XRT data in 2014 and 2015. The flux of M82 X-1 increased by a factor of 2-3 from 2014 to 2015. Most of the data in 2015 show greater dominance of hard component than those of 2014. Due to moderate signal-to-noise ratio, we only fit each spectrum with power-law and disk blackbody model separately. The data in 2014 are better fitted with `powerlaw` model based on the value of reduced-chi squared. On the other hand, both `powerlaw` and `diskbb` models showed comparable reduced chi-squared value for the data in 2015. We found that the range of spectral index for 2014 data is $1.65 < \Gamma < 2.08$ and for 2015 data is $1.02 < \Gamma < 1.95$ from the `powerlaw` model, resembling the range for that of black hole binary system at low mass accretion rate. We obtained higher innermost disk temperature from the disk blackbody model, 1.20 keV $< T_{\rm in} <$ 3.63 keV, compared to that of black hole binary system in the thermal state. The calculated innermost radius of the disk, $R_{\rm in}$, varies between 0.99 to 4.89 $R_{\rm S}$ assuming 10 $M_\odot$ black hole which indicates that the spectral state is not in thermal dominant state but rather we suspect that M82 X-1 exhibits greater mass accretion rate than that of the thermal dominant state.
Keywords: M82 X-1, black hole accretion, spectral evolution.

1 Introduction

Ultraluminous X-ray sources (ULXs) are defined as point-like X-ray sources located outside the center of galaxies with luminosities exceeding 3×10^{39} erg s^{-1} [3]. These apparent luminosities, assuming isotropic emission, are above the Eddington limit of a 20 $M_\odot$ black hole. Despite the fact of being very luminous, ULXs cannot be explained by the collections of many sources with each luminosity less than the Eddington limit because many ULXs exhibit significant time variabilities [6]. One reasonable assumption is that the ULXs are single compact objects powered by accretion. If the Eddington limit applies to these objects, then ULXs is one of the best candidates for an intermediate-mass black hole (IMBH). Alternatively, ULXs may represent a class of super-Eddington emitters and probes to a regime in

*e-mail: izatulhafizah@students.itb.ac.id
**e-mail: kiki@as.itb.ac.id

which the accretion rate is much higher than that seen in Galactic black hole binaries (e.g. Brightman et al. 2020)[2]. The reality is probably more complex, with some ULXs being possibly more massive than stellar-mass black holes ($M_{ULX} > 100\ M_\odot$), but still experiencing the super-Eddington accretion rate. M82 X-1 is one of the brightest ULXs, which can reach $L_X \sim 10^{41}$ erg s^{-1} at the distance of 3.4 Mpc [4]. Its high luminosity, makes it one of the best intermediate-mass black-hole (IMBH) candidates, assuming Eddington-limited accretion. Detection of twin-peaked QPOs at 3.3 and 5.1Hz lead to a mass estimate using scaling laws between the QPOs frequencies and mass of $428 \pm 105\ M_\odot$ [8]. However, the mass estimates for M82 X-1 using different methods vary considerably and have large uncertainties. Therefore its status as an IMBH has not been firmly established. For instance, modelling of the accretion disk emission shows that M82 X-1 can be explained by a $\sim 30\ M_\odot$ stellar remnant black hole radiating at several times its Eddington limit [7].

Here, we report our X-ray spectral analysis using *Swift*/XRT observations. We aim to investigate the variability of the X-ray spectra and estimate the mass of the black hole in M82 X-1. In Section 2, we describe the X-ray data which we used in the this study, including the details of data reduction and fitting methods. In Section 3, fitting results are presented, together with their implications on the black hole mass. We conclude our work in Section 4.

2 Data and Methods

2.1 Data

We use *Swift*/XRT data for our study due to its frequent observations during 2014 and 2015. We chose 30 observation data which represent different intensity, consisting of 12 observations from 2014 and 18 observations from 2015. For 2014 data, only the observation data which have an exposure time > 5000 seconds are used. Low exposure data are not used due to their low number of counts which render insufficient statistics. The observations in 2014 represent low intensity data. Reported by Brightman et al. (2016)[1], the flux of M82 X-1 is found to increase during May to July in 2015. Therefore, we have added 18 observations data during May to July 2015. Table 1 provides a description of the observational data.

2.2 Data Reduction and Fitting Methods

We extracted the source counts from a circular region with a radius of ~ 49" and the background events were extracted from a nearby circular region of the same size (Figure 1). The hardness ratio is defined as the ratio between the count rate in 1.5-10 keV to that of 0.3-1.5 keV.

Figure 2 shows the 0.3-10 keV count rate of each observation (light curve) and the corresponding hardness ratio. The hardness ratio shows that 2015 data are more dominated by the hard X-ray component than those of 2014. On the bottom panel, the light curve shows that the flux increases of about 2 to 3 times from 2014 to 2015. This shows a spectral evolution that is different from that of Galactic black hole binary, in which the softer component increases as the source brightened. The spectra were grouped with a minimum of one count per bin. The C-statistic was used for spectral fitting. The spectra were fitted in the range 0.3–10 keV with two different model, `powerlaw` and disk black body (`diskbb`).

3 Spectral Analysis

Not all of the data can be fitted by single `powerlaw` or `diskbb` model. Table 2 shows percentage of well-fitted data to `powerlaw` or `diskbb` model for 2014 and 2015 data. The fitting

Table 1. Observational data

Observation ID	Observation Date	Exposure Time (second)
33123024	2014-01-31 00:54:59	13456.233
33123006	2014-01-25 08:41:59	12286.459
33123023	2014-01-31 00:42:13	11194.395
33124001	2014-01-23 00:48:11	9974.064
33123025	2014-01-31 18:24:57	9812.241
33123017	2014-01-30 02:09:34	9177.898
33123063	2014-02-18 00:33:59	8607.612
33123068	2014-02-22 00:34:58	7932.295
33123018	2014-01-30 16:33:04	6423.361
33123053	2014-02-10 08:37:59	5309.887
33123035	2014-02-04 10:10:59	5293.636
33123032	2014-02-02 21:40:59	5003.257
92202007	2015-05-05 03:43:59	2020.883
92202008	2015-05-10 08:21:59	1965.717
92202009	2015-05-15 06:17:59	1840.302
92202010	2015-05-20 15:35:59	1672.356
92202011	2015-05-25 16:59:59	1747.573
92202012	2015-05-30 15:02:22	1144.226
92202013	2015-06-04 02:00:59	2194.543
92202014	2015-06-16 13:01:59	1970.731
81460001	2015-06-20 17:28:58	1648.556
92202015	2015-06-21 12:32:59	2161.162
92202016	2015-06-26 06:00:59	1854.772
92202017	2015-07-01 07:02:58	1852.886
92202018	2015-07-06 17:51:58	1766.21
92202019	2015-07-11 06:25:57	2055.964
81460002	2015-07-11 23:58:58	1875.44
92202020	2015-07-16 20:35:57	2013.352
92202021	2015-07-21 04:02:58	2124.125
92202022	2015-07-26 10:11:58	2052.384

Table 2. Well-fitted data percentage

Model	2014	2015
`powerlaw`	66.6%	83.3%
`diskbb`	25%	77.7%

result is plotted in Figure 3. `Powerlaw` fittings show the range of spectral index, 1.02 to 2.08, which is similar to the range of spectral index for that of black hole binary system at low mass accretion rate. The range of $T_{\rm in}$ values obtained from `diskbb` model is 1.20 to 3.63 keV, which is higher than typical innermost temperatures of thermal state in non-rotating BHB, but 35% of the well-fitted data gives a comparable value compared with temperatures of maximally rotating BHB.

We calculated the X-ray luminosity of M82 X-1 using the flux from fitting and assuming the distance of M82 X-1 is 3.4 Mpc. Using the innermost temperature from fitting result and luminosity that has been calculated before, we calculated the innermost radius of the disk,

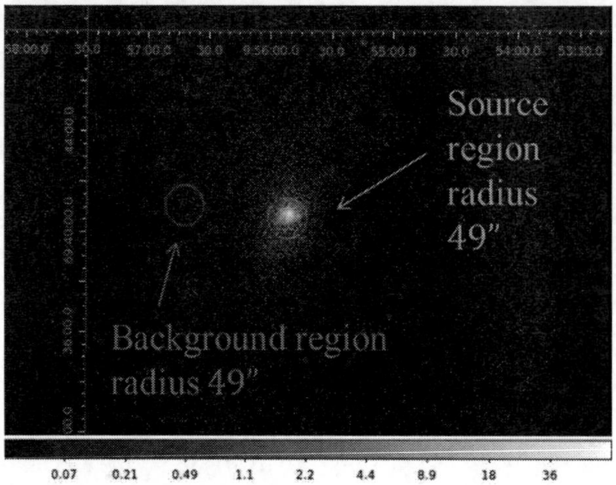

Figure 1. *Swift*/XRT 0.3-10 keV image of M82 X-1. The green circle shows the source extraction region and the magenta circle for background region, both have a radius of 49".

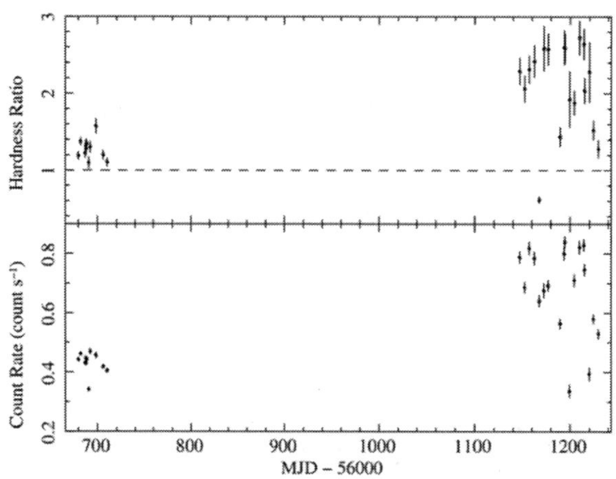

Figure 2. hardness ratio (top panel) and light curve (bottom panel).

$R_{\rm in}$, assuming the radiation is dominated by black body radiation. We used $\kappa \sim 1.7$ [5] which is the spectral hardening factor, and $\xi = 0.412$ to correct $R_{\rm in}$. Then the mass is calculated using $M/M_\odot = R_{\rm in}/8.86\beta$ with a β value 1 for non-rotating and 1/6 for maximally rotating black hole.

Figure 4 is temperature-luminosity plot of 2014 observations (black) and 2015 observations (red). Both 2014 and 2015 data do not show a $L \propto T^4$ relation which is commonly found in the thermal state of black hole binary (BHB). The data show a $L \propto T^{0.67}$ relation for 2014 and $L \propto T^{1.07}$ relation for 2015. It means the data used in this study do not follow the trend of the thermal state of BHB. This is confirmed by the radius-flux plot (Figure 5) which gives a non-constant innermost radius values. Spectrum in the thermal state will produce a more or less constant radius values. Figure 6 shows plot of radius-mass, in which a non-rotating black

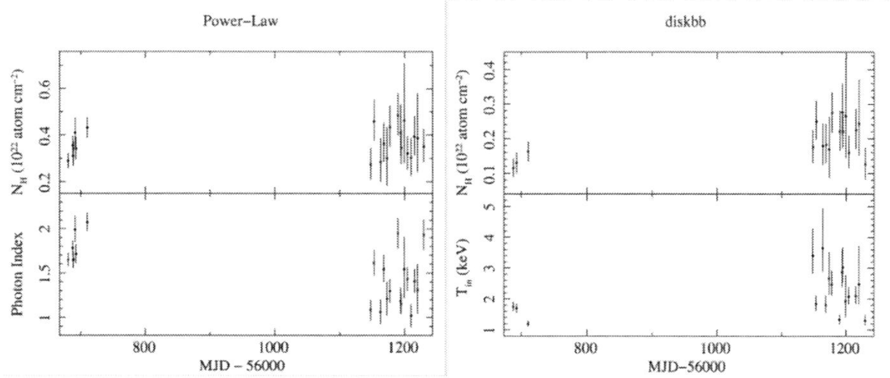

Figure 3. hardness ratio (top panel) and light curve (bottom panel).

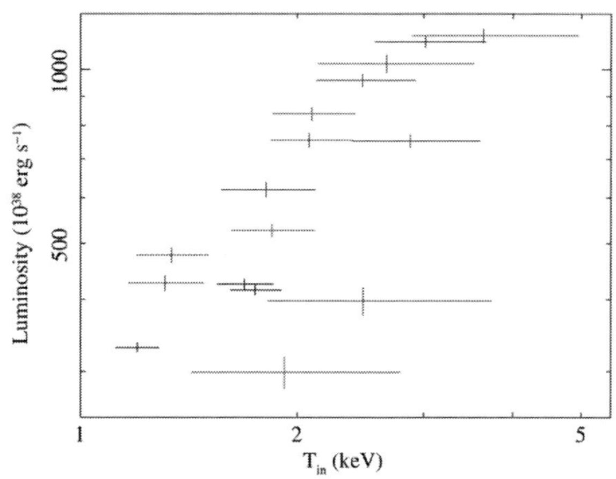

Figure 4. Temperature-luminosity plot, black for 2014 data and red for 2015 data.

hole is assumed in the data points without star symbol while maximally rotating black hole is assumed for data with star symbol. This plot shows that the obtained black hole masses varied due to the non-constant values of innermost disk radius, which cannot be physical, but instead reveal the limitation of mass estimate based on spectral fitting when the standard disk assumption cannot be justified.

4 Conclusion

We have used *Swift*/XRT data to show that the spectral variability of M82 X-1 does not resemble that of BHB. The increasing domination of hard component of the spectra as the intensity increase does not necessarily mean that the non-thermal component predominates the spectra, but may imply a higher inner disk temperature which support supercritical accretion, assuming a constant spin of the black hole. The choice of spectrum used to determine black hole mass, must be done carefully. Despite this limitation, all data used in this study indicate a black hole mass less than 100 solar mass. Based on the obtained range of black hole mass, we conclude that M82 X-1 requires a higher accretion rate to explain the different spectrum

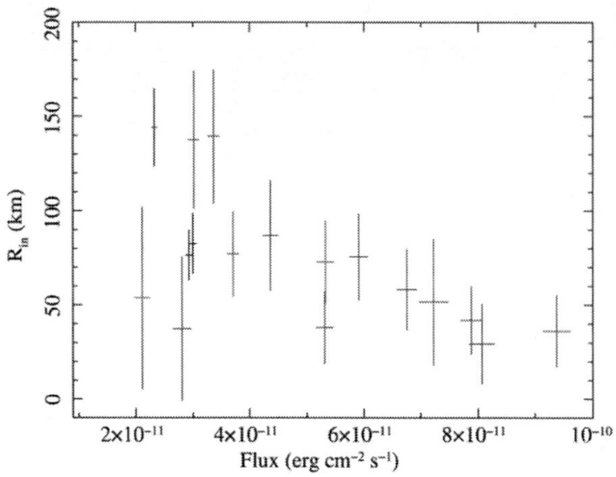

Figure 5. Flux-innermost radius plot, black for 2014 data and red for 2015 data.

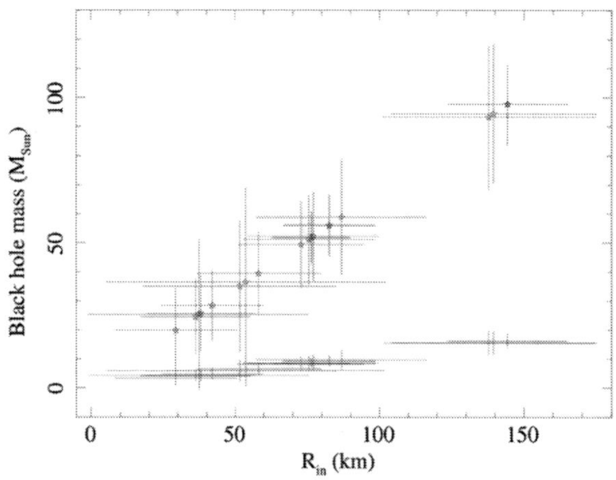

Figure 6. Radius-black hole mass plot, black for 2014 data and red for 2015 data. Star symbol for maximally rotating black hole.

transitions compared to BHB at sub-Eddington accretion rates. We therefore conclude that M82 X-1 is a super-Eddington accretor.

References

[1] Brightman, M., Harrison, F. A., Barret, D., et al., ApJ **829**, 28 (2016)
[2] Brightman, M., Walton, D. J., Xu, Y., et al., ApJ **889**, 71 (2020)
[3] Feng, H. & Soria, R., NewAR **55**, 166 (2011)
[4] Foley, R. J., Fox, O. D., McCully, C., et al., MNRAS **443**, 2887 (2014)
[5] Kato, S., Fukue, J., & Mineshige, S., *Black-Hole Accretion Disks* (Kyoto University Press, Kyoto, 2008) 60
[6] Makishima, K., Kubota, A., Mizuno, T., et al., ApJ **535**, 632 (2000)

[7] Okajima, T., Ebisawa, K., & Kawaguchi, T., ApJL **652**, L105 (2006)
[8] Pasham, D. R., Strohmayer, T. E., & Mushotzky, R. F., Nature **513**, 74 (2014)
[9] Poutanen, J., Lipunova, G., Fabrika, S., et al., MNRAS **377**, 1187 (2007)

EPJ Web of Conferences **240**, 07006 (2020) https://doi.org/10.1051/epjconf/202024007006
SEAAN Meeting 2019

Transverse Energy Density in High Energy Heavy Ion Collisions

Kian Hwee Lim[1,*], *Aik Hui* Chan[1], and *Choo Hiap* Oh[1]

[1]Department of Physics, National University of Singapore, 2 Science Drive 3, Singapore 117551

Abstract. A phenomenological model describing the transverse energy distribution (E_T) of nuclear collisions is first studied in detail by fitting it on E_T data for O-Pb collisions at $\sqrt{s_{NN}}$ = 200 GeV per nucleon obtained from the NA35 collaboration. Next, the model is used to fit the E_T data for Pb-Pb collisions at LHC energies of $\sqrt{s_{NN}}$ = 2.76 TeV per nucleon obtained from the ATLAS collaboration. From the fits, we determine an upper bound for the energy density for Pb-Pb collisions at LHC energies of $\sqrt{s_{NN}}$ = 2.76 TeV per nucleon.

1 Introduction

According to the Standard Model, hadronic particles are bound states of quarks. The quarks in the hadronic particles interact via the strong force and are bound by the strong force. Although currently, no isolated quark has been observed, results from deep inelastic scattering show that the quark model is highly viable [1].

Current theories say that as temperature increases, the strong force weakens. The opposite is also true. This is known as Asymptotic freedom in QCD. The implication of Asymptotic freedom is this: at high enough temperatures, we might have a possible new state of matter, where the quarks exist not as bound states, but as free particles. This new state of matter has been termed "Quark Gluon Plasma". Going from hadronic particles to Quark Gluon Plasma at high temperatures is known as Deconfinement, whereas going from Quark Gluon Plasma to hadronic particles at low temperatures is known as Hadronisation. Both Hadronisation and Deconfinement can be throught of as phase transitions [2]. Current theoretical calculations using Lattice QCD show that this phase transition occurs at a critical energy density of 0.7 ± 0.2 GeV fm^{-3} [3] and at a temperature of 140-190 MeV [4].

According to Big Bang Cosmology, the early universe was both extremely hot and extremely dense. This means that Quarks and Gluons probably existed as Quark Gluon Plasma in the early universe. Then, as the universe expanded and cooled, the Quark Gluon Plasma underwent Hadronisation to give us the hadrons that we see today. While various people have tried to model this process of Hadronisation, such as in [10], this process of Hadronisation is still under active investigation. Clearly, a better understanding of Quark Gluon Plasma would help us achieve a better understanding of Hadronisation. Thus, if we are able to make Quark Gluon Plasma in the lab, we will be able to better study the Hadronisation process.

From as early as the 1980s, the collision of heavy ions at as high energies as possible was seen as a viable way to experimentally replicate the hot and dense conditions of the

*e-mail: kianhwee_lim@u.nus.edu

early universe, which is needed for the formation of Quark Gluon Plasma. Such heavy ion collisions were first done in colliders like the RHIC (Relativistic Heavy Ion Collider), and are done today in experiments like ALICE (A Large Ion Collider Experiment) in CERN. There are certain theoretical signposts that tell us if we have indeed created Quark Gluon Plasma in our heavy ion collision. Firstly, we need to first check if the energy density in the heavy ion collision is greater than the theoretical critical energy density. Next, we need to check for other theoretical signs of Quark Gluon Plasma, such as Strangeness enhancement [5], amongst many others.

The remainder of this paper is to explore the first point above with respect to Pb-Pb collisions at LHC energies of $\sqrt{s_{NN}}$ = 2.76 TeV per nucleon obtained from the ATLAS collaboration. A phenomenological model will be used for this purpose.

1.1 Purpose of this paper

We want to be determine the energy density in Pb-Pb collisions at LHC energies of $\sqrt{s_{NN}}$ = 2.76 TeV per nucleon obtained from the ATLAS collaboration. In other words, from the collider data, we want to extract the energy density. There are no "first principles" (i.e QCD) model to do this yet, thus the problem has to be treated phenomenologically.

The collider data that we want to study is shown below in figure 1. In this report, we show that the upper bound for the energy density for the collision data in figure 1 is 0.006 TeV. Future work will be done to determine the exact energy density.

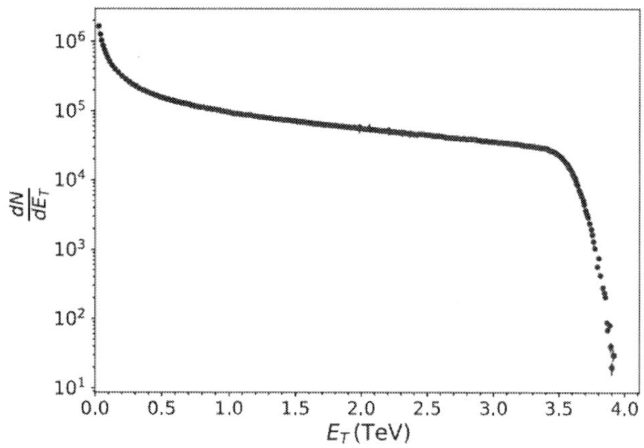

Figure 1. The transverse energy distribition for Pb-Pb collisions at centre of mass energy 2.76 TeV per nucleon. The data here is in the pseudorapidity range of $3.2 < |\eta| < 4.9$, and is taken from the ATLAS collaboration [6].

1.2 Approach

As mentioned, a phenomenological approach to finding the energy density will be taken. Thus, we will first construct a phenomenological model to model heavy ion collisions. The model will take in as input the transverse energy distribition data as shown in figure 1, and produce as the output a number that can be reasonably interpreted as the energy density of the heavy ion collision. Next, we will test the model we constructed on an old and well-investigated set of data, shown in figure 2, to try and reproduce past results. Lastly, if the previous step is successful, we will test the model on the real data shown in figure 1.

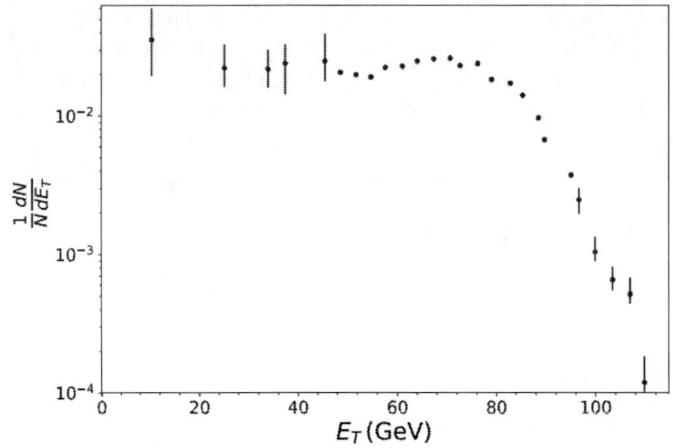

Figure 2. The data for ^{16}O on ^{208}Pb at energies of $\sqrt{s_{NN}}$ = 200 GeV per nucleon in pseudorapidity range $2.2 < \eta < 3.8$. Data is from the NA35 collaboration [9], and has already been studied in detail in [7].

2 The Phenomenological model considered

As was mentioned previously, we will first need to find a suitable phenomenological model. We will adapt a geometrical model used before in the past to analyse heavy ion collisions [7],[8]. A rough schematic of the geometrical model is shown in figure 3. There are three key

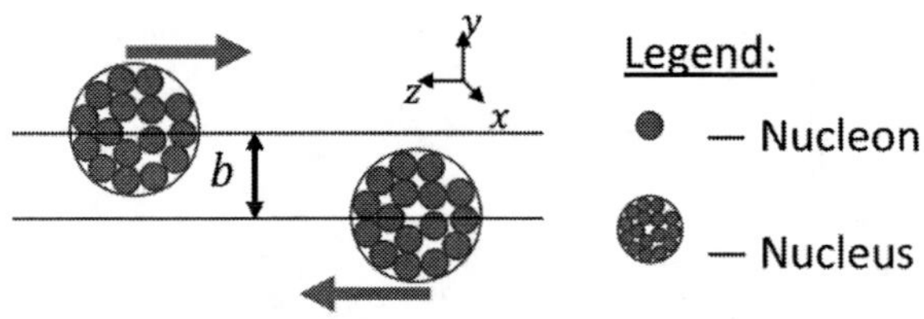

Figure 3. Geometrical model considered. The two colliding nuclei are considered as spheres colliding at a given impact parameter b.

ideas behind this geometrical model. They are:

1. Treat nuclei-nuclei collisions as many individual nucleon-nucleon collisions
2. Let each nucleon-nucleon collision collision have a transverse energy distribution $\rho(E_T)$, with mean transverse energy per collision ϵ
3. Let the number of nucleon-nucleon collisions n take a Poisson distribution about the mean $\bar{n}(b)$

From these three key ideas, we can derive the following equation for transverse energy density:

$$\frac{1}{N}\frac{dN}{dE_T} = \left\langle \sum_{k=0}^{\infty} \rho^k(E_T) \frac{\bar{n}(b)^k}{k!} e^{-\bar{n}(b)} \right\rangle \quad (1)$$

where in equation 1, $\rho^k(E_T) \equiv \int \prod_{i=1}^{k} \rho(E_i)\delta(E_T - \sum_{a=1}^{k} E_a)$ denotes a k-fold convolution of ρ, and the angular brackets $\langle\,\rangle$ denote an average over the impact parameter b. Equation 1 has 1 free parameter, which is ϵ. Thus, we can vary the value of ϵ to best fit the data in figures 1 and 2. The value of ϵ that best fits the data is the value that we want. Since ϵ is the mean transverse energy per nucleon-nucleon collision, ϵ can be interpreted as the energy density during the Pb-Pb collision. Thus, the energy density during the nuclei-nuclei collision is determined.

Now, we see that to evaluate the RHS of equation 1, we need to calculate the mean number of collisions at a given impact parameter, $\bar{n}(b)$. This is the hardest part of evaluating equation 1, and this is why this model is known as a geometrical model.

2.1 Calculating $\bar{n}(b)$

Calculating $\bar{n}(b)$ is the key to evaluating equation 1. To do so, we firstly need to define the nuclear profile of the two nuclei. The Wood-Saxon nuclear density profile is used to get the results below, but it has been shown in [7] that the results obtained are pretty independent on the nuclear density profile used. This will be verified in a future work.

Using the Wood-Saxon nuclear density profile, we have

$$R = 1.19A^{1/3} - 1.61A^{-1/3} \tag{2}$$

$$\rho(r) \propto (1 + \exp(r - R)/s)^{-1} \tag{3}$$

where $\rho(r)$ is normalised such that $\iiint \rho(r)\,d\mathbf{r} = A$, where A is the atomic number of the nuclei. Here, s which is the surface thickness of the nucleus is taken to be 0.54 fm^{-3}.

Now, consider the case of two colliding nuclei A and B with nuclear density profiles $\rho_A(r)$ and $\rho_B(r)$ respectively. Also, let the nucleon-nucleon inelastic scattering cross section be σ_{nn}. By geometrical considerations, we would have:

$$\bar{n}(b) = \iint dxdy\,\sigma_{nn} \int dz_A\,\rho_A\Big(\sqrt{x^2 + y^2 + z_A^2}\Big) \int dz_B\,\rho_A\Big(\sqrt{x^2 + (y-b)^2 + z_B^2}\Big) \tag{4}$$

where b is the impact parameter of the two nuclei. The x-y integration is done over the overlap region of the two geometrical cross sections, as shown in figure 4.

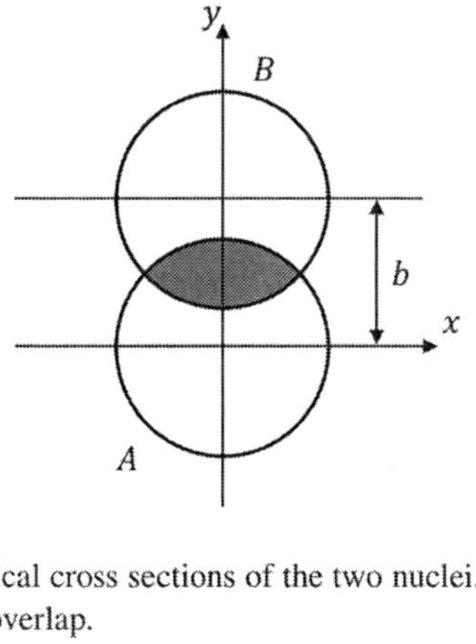

Figure 4. Overlap of the geometrical cross sections of the two nuclei. The z axis points into the page. $\bar{n}(b)$ can be determined from this overlap.

Using the Pb-Pb collision as an example, equation 4 would look like:

$$\bar{n}(b) = \int_0^{x_I} 2\,dx \int_{b-\sqrt{R^2-x^2}}^{\sqrt{R^2-x^2}} dy\,\sigma_{nn} \int_0^{z_A} dz_A\,\rho_A\Big(\sqrt{x^2+y^2+z_A^2}\Big) \int_0^{z_B} dz_B\,\rho_A\Big(\sqrt{x^2+(y-b)^2+z_B^2}\Big)$$

where R is the radius of the Pb nuclei in the Wood-Saxon nuclear density profile, and

$$x_I = \sqrt{R^2 - \frac{b^2}{4}}$$
$$z_A = \sqrt{R^2 - x^2 - y^2}$$
$$z_B = \sqrt{R^2 - x^2 - (y-b)^2}$$

When the two nuclei have different radii, more care has to be done when using equation 4. In this case, these three cases have to be considered separately:

- $0 \le b < R_A - R_B$
- $R_A - R_B \le b < \sqrt{R_A^2 - R_B^2}$
- $\sqrt{R_A^2 - R_B^2} \le b < R_A + R_B$

where R_A is the radius of nuclei A and R_B is the radius of nuclei B, and we assume $R_A > R_B$.

3 Validating the geometrical model

Now that we have explained the geometrical model above, we fit equation 1 to the data in figure 2 to determine the value of ϵ, which will be interpreted as the energy density. Our results are shown figure 5.

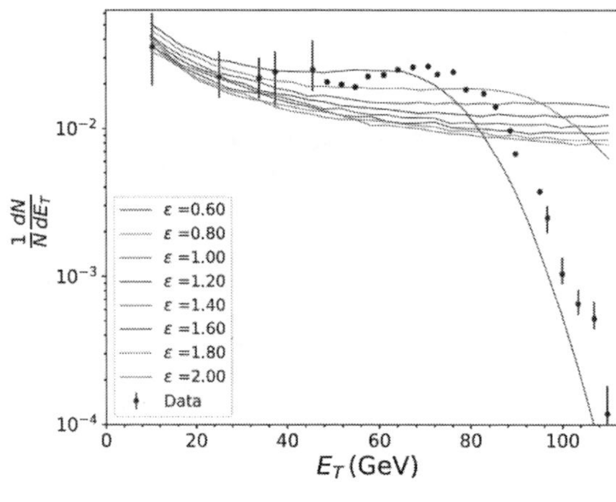

Figure 5. Equation 1 fitted with various values of ϵ to the data in figure 2.

Let us first examine the graphs of $\epsilon = 0.6$ GeV and $\epsilon = 0.8$ GeV. The graphs of $\epsilon = 0.6$ GeV and $\epsilon = 0.8$ GeV are unable to fit the data as the locations of the "shoulder" are too far to the left and too far to the right respectively. Thus, we infer that the value of ϵ that best fits the data is between 0.6 GeV and 0.8 GeV.

This is consistent with the results obtained from [7], who calculated the energy density of this set of data to be 0.79 GeV.

Another thing to note would be that as the value of ϵ increases, the location of the "shoulder" shifts rightwards. Thus, for the other values of ϵ plotted, the location of the "shoulder" is too far to the right to be seen on this diagram.

4 Preliminary results for Pb-Pb collision

Since our geometrical model could successively reproduce past results, we apply it to the our data in figure 1. Our results are shown in figure 6.

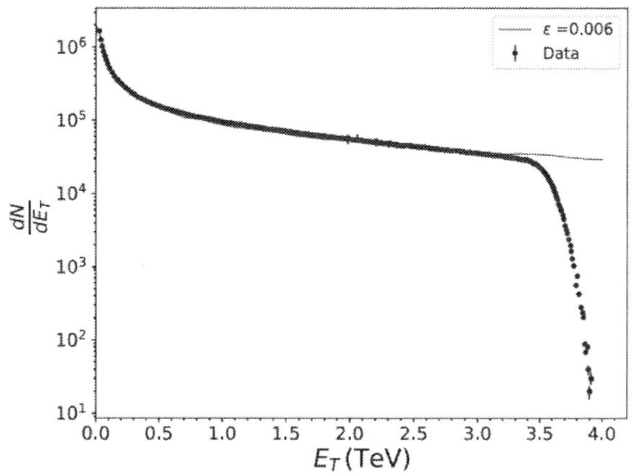

Figure 6. The fit of equation 1 with $\epsilon = 0.006$ TeV to the data in figure 1.

From figure 6, we see that the location of the "shoulder" as predicted by equation 1 with $\epsilon = 0.006$ TeV is too far right to be shown on the graph. Thus, we conclude that for the data in figure 1, the energy density must be lower that $\epsilon = 0.006$ TeV.

5 Conclusion

According to the predictions of both the Standard Model and Big Bang cosmology, Quark Gluon Plasma is possibly in the early universe. Then, as the universe cooled, the Quark Gluon Plasma hadronised to form hadrons. This provides ample motivation for studying high energy heavy ion collisions, which tries to mimic the conditions of the early universe to produce Quark Gluon Plasma. To determine the existence of Quark Gluon Plasma during such high energy heavy ion collisions, we need to check if the energy density reached during the collision exceeds the theoretical critical minimum value required for Hadronic matter to undergo a phase transition to Quark Gluon Plasma.

The problem we want to solve in this paper is this: from the heavy ion collision data in figure 1, we want to determine the energy density reached during the collision. To that end, a geometrical model is used for phenomenological calcuations. Preliminary calculations show that the energy density is lower than 0.006 TeV. Future work will have to be done to determine exactly what the energy density is.

References

[1] D.J Griffiths, *Introduction to Elementary Particles* (Wiley-VCH Verlag, 2014)
[2] P.Braun-Munzinger and J.Stachel, Nature **448**, 302 (2007)
[3] F.Karsch, E. Laermann, and A.peikert, Nuclear Physics B **605**, 579 (2001)
[4] Y.Aoki, S. Borsanyi, et al., Journal of High Energy Physics **2009**, 088 (2009)
[5] J.Rafelski and B. Muller, Physical Review Letters **48**, 1066 (1982)
[6] T.Bold, in *Physics at the LHC 2011* (The ATLAS collaboration, 2011)
[7] A.Jackson and H.Boggild, Nuclear Physics A **470**, 669 (1987)
[8] A.H Chan and C.H Chew, II Nuovo Cimento A **104**, 81 (1991)
[9] A.Bamberger et al., preprint CERN/EP 86-194 (1996)
[10] G.Baym, Nucl. Phys. **A447**, 463C-473C (1986).

A Study of Ethnoastronomy in Ciptagelar Civilizations, Sukabumi Regency, West Java, Indonesia: Relation of Full Moon Phase with Javanese Calendar and Hijriyah Calendar

Dewi Muthia[1,*], *Aprilia*[1]

[1]Astronomy Division, Faculty of Mathematics and Natural Sciences, Institut Teknologi Bandung, Jl. Ganesa 10, Bandung, 40132, Indonesia

Abstract. Astronomy has been practiced for as long as humans have been looking at the sky. Early civilizations used several celestial objects to do their daily activities, like doing a traditional ceremony. Ciptagelar Village is located at Sukabumi Regency, West Java, Indonesia. It has been established for almost 650 years. Ciptagelar villagers consider that a full Moon phase is a special and sacred thing. They have a welcoming ceremony for it, named Mapag Purnama. We study this activity in the way of Ethnoastronomy. The methods that we used are interviewing the villagers, including the customary leader of Ciptagelar Village, and doing a literature study. Ceremony of Mapag Purnama is held every 13rd to the night of 14th of the Moon Javanese Calendar. We use three references for choosing the date of full Moon phase, namely Fred Espenak, NASA, and Stellarium application. This date is then converted from Solar Calendar to Javanese and Hijriyah. Because Javanese Calendar is adopted from Hijriyah Calendar, both of two will have the same dates. We use 120-years correction for a difference cycle and number of leap years. However, after 2010 CE this correction is no longer valid. We find that the Javanese Calendar is one day slower than the Hijriyah Calendar.

1 Introduction

Indonesia is an archipelago. It has many islands as well as ethnicity and culture. In this modern era, it turns out that there are still a number of ethnic groups whom uphold their customs until now. One of them is the Ciptagelar Civilization which we study in this paper. They are located in Sukabumi Regency, West Java, Indonesia.

We conduct this research because of three motivations. First, we like a science that has a direct relation to the civilizations. We choose to study the ethnoastronomy in this way. "Ethnoastronomy is the study of astronomies of different races, people or ancient cultures which includes folk or indigenous astronomies, calendars, celestial lore, sky mythology and related ritual, cosmological concepts and celestial-related traditions of all people and cultures." [1]. We try to understand how science affects their lives. Second, Indonesia is a

* Corresponding author: mthdw788@gmail.com

country that is suitable for doing an ethnoastronomy research. So many islands and differences in each ethnicity we can observe. Indonesia has an abundance of nature and culture.

Civilizations use many signs in nature to assist their daily activities [4, 5, 6, 7, 8, 11]. One of them is related to astronomy, i.e. they have seen the sky and used several celestial objects to help them know about time to work, to farm, etc. Each ethnicity has a unique traditional ceremony related to observing the sky. Why then do we choose to study the Ciptagelar civilization than other civilizations in Indonesia? This question asks whether the third thing that motivates us. That is because Ciptagelar civilization is an unique civilization than others. What makes them unique? As we knew about customary civilizations, they tend to be backward, closed, and rarely want to accept technology, but this is different with Ciptagelar civilizations. Indeed, they are a customary civilization, but they live in an advanced village that is very open to the outside world and technology savvy.

The Ciptagelar villagers have several kinds of ceremonies. A few ceremonies are related to astronomy knowledge. One of them is *Malam Opat Belasan*. This name is taken from the Sundanese language and means a ceremony that held in every 13rd to the night of 14th in Javanese Lunar Calendar. This ceremony is conducted to welcome the full Moon phase. Thus, in Sundanese this ceremony also called *Mapag Purnama*.

2 Method

This research uses some methods to collect the data. First, we conduct a direct interview with the villagers and the customary leader of the Ciptagelar Village. Second, we conduct a literature study. To do the interview, we have to prepare the three steps, i.e. the preparation, the interview, and the post interview [9, 10]. For the preparation, we have to contact the villagers for asking permission related to their willingness to be interviewed. Then, if they are willing, we propose our arrival date to conduct the interview process. Then, we have to prepare the questions that will be asked to them in the interview. We also have to make sure our belongings that will be brought to their village and funds needed during the research.

The next step is about the interview. In this step, we just have to do well the interview process with questions that already prepared before. We also documented it during the interview. The last step is the post interview. In this step, we thanked the interviewees for their willingness to be interviewed. We also ask them about a possibility to conduct a further interview if needed.

3 The Ceremony of *Opat Belasan*

This ceremony is intended to welcome the full Moon phase. This ceremony aims to preserve ancestral traditions and mysteries. In Indonesian, it is referred to as *wangsit* from the ancestors to the customary leader of Ciptagelar Village. This ceremony enlivened by many art performances, such as *Angling Dogdog Lojor, Topeng Kolot, Wayang Golek Rurukan, Jipeng,* and *Degung Gonjringan*. The interesting thing about the art performance of *Wayang Golek Rurukan* is the story of the ancestors [13, 16, 17, 18]. The Ciptagelar Villagers carried out this ceremony in the customary building, i.e. *Imah Gede*. This building is also the house of the customary leader, named Abah Ugi.

4 The Javanese Calendar System

The Javanese Calendar has been used since 1600s CE. The Javanese Calendar is the result of an assimilation of the Indian Calendar, named Saka Calendar and Hijriyah (Islamic) Calendar. In the Javanese Calendar, the year numbering continues the year number of the Saka Calendar, the date and month are derived from the Hijriyah Calendar. Thus, in 1633 CE, the Javanese Calendar has a year number of 1555 Java instead of 1043 Hijriyah (H) [14].

The Javanese Calendar has its own system of cycles. The two longest unit of time in Javanese Calendar are *windu* and *lambang*. Both of them have the same duration, i.e. eight years, but they have a different cycle. The total cycle of *lambang* is 16 years that is divided into two kinds of *lambang*, i.e. *langkir* and *kulawu*. For *windu*, it has a total cycle of 32 years that is divided into four kinds of *windu*, i.e. *adi, kuntara, sengara*, and *sancaya*.

From *lambang* and *windu*, there are eight kinds of year, i.e. *Alip* (354 days), *Ehe* (355 days), *Jimawal* (354 days), *Je* (354 days), *Dal* (355 days), *Be* (354 days), *Wawu* (354 days), and *Jimakir* (355 days). The Javanese have two kinds of calendar systems, i.e. the *Pranata Mangsa* (Solar Calendar) and the Lunar Calendar. This eight-year cycle is only for the Javanese Lunar Calendar. The 2nd, 5th, and 8th year are categorized as a leap year where a year consists of 355 days [3]. The additional day for the leap year is added to the 12th month. The name and duration for each month on the Moon Javanese Calendar can be seen in the Table 1 below.

Table 1. The Name of Months in the Moon Javanese Calendar and their Duration.

No.	Name of Months	Duration (days)
1.	*Sura*	30
2.	*Sapar*	29
3.	*Mulud*	30
4.	*Bakda Mulud*	29
5.	*Jumadilawal*	30
6.	*Jumadilakir*	29
7.	*Rejeb*	30
8.	*Ruwah*	29
9.	*Pasa*	30
10.	*Sawal*	29
11.	*Sela*	30
12.	*Besar*	29 or 30

In the Javanese Calendar, there are the terms *Pancawara* and *Saptawara*. Both of them are unit of time for the week. *Pancawara* consists of five days, namely *Legi, Pahing, Pon, Wage*, and *Kliwon*, while *Saptawara* consists of seven days, namely *Ahad, Senin, Selasa, Rabu, Kamis, Jum'at* and *Sabtu*. Both run concurrently and are known as *Pasaran* in Javanese. Besides *Pancawara* and *Saptawara*, there are still six other kinds of *Pasaran*, i.e. *Dwiwara* (2 days), *Triwara* (3 days), *Caturwara* (4 days), *Sadwara* (6 days), *Astawara* (8 days), and *Sangawara* (9 days). Also, in the Javanese Calendar, there is another unit of time that is a combination of *Pancawara* and *Saptawara* namely *Wuku*. Each type of *Wuku* will change every week which consists of seven days.

The first month in the Javanese Calendar is *Sura*, thus first *Sura* is the beginning of the Javanese Calendar year. To decide this beginning of the year, the Javanese Calendar follows the Hijriyah Calendar year. Thus, first *Sura* coincides with first *Muharram* (the first month in Hijriyah Calendar).

5 The Determination of the Date of Occurrence of the Full Moon Phase and Its Relation to the Ceremony of *Malam Opat Belasan*

Ciptagelar villagers have their own perception about full Moon phase. Based on their opinion, a full Moon phase is something sacred. We know that Ciptagelar villagers have a ceremony that relates to the full Moon phase, namely *Malam Opat Belasan*. It is held every the 13rd to the night of 14th every month in the Moon Javanese Calendar.

When we interviewed one of the Ciptagelar villagers named Mr. Yoyo, he told that the ceremony of *Malam Opat Belasan* has two rituals. The first ritual is held at the night of the full Moon phase and the second is held after that full Moon phase. Also, he said that determining the date for the occurrence of the full Moon phase was based on the first experience of their ancestors who saw the full Moon phase in 1368 CE. That year, their ancestors saw the full Moon phase on the 13rd to the night of 14th in the Javanese Lunar Calendar. Accordingly, this date is set as the date of the full Moon phase in the Javanese Calendar. Until now, this date has been used as a celebration for the ceremony of *Malam Opat Belasan*.

To confirm Mr. Yoyo's explanation, we trace back the date and check whether it is true or not. To do so, we convert that date from the Western Calendar (in this case the Julian Calendar) to the Javanese Lunar Calendar. There are two ways to perform the conversion. First, we convert directly from the Gregorian Solar Calendar to the Hijriyah Calendar using the Hisab Urfi Calculation. Then, we use the year number in the Hijriyah calendar to obtain the year number in the Javanese Lunar Calendar. The other method is to convert the date from the Gregorian Solar Calendar to the Hijriyah Calendar then change the year number to the Javanese Lunar Calendar, and then apply the 120-year correction between the Javanese and Hijriyah Calendar.

6 *Hisab Urfi* Islamic Dating System

There are many types of Islamic Dating System, such as Hisab Hakiki, and Imkan Rukyat, but for this conversion, we will use the Hisab Urfi dating system. This dating system uses the Synodic cycle (based on observations) of Moon on average, i.e. 29 days, 12 hours, 44 minutes, and 2.8 seconds in every month as a reference calculation. This value is a time duration from one conjunction to another. Initially, the dating system began when the *hijrah* (moving from one place to another) event of the *Prophet Muhammad SAW* from Mecca to Medina that occurred on 15th or 16th of July 622 CE.

The determination of the beginning of the month is based on *hilal* (thin crescent Moon) observation that can be seen after the Sunset. Hisab Urfi Islamic dating system is based on calculation. It means that it tends to predict rather than observe in determining something (for example, an astronomical phenomenon). A year consists of 12 months and each month consists of 29 days (for even months) or 30 days (for odd months). The names of the 12 months are *Muharram, Safar, Rabi'ul Awal, Rabi'ul Akhir, Jumadil Awal, Jumadil Akhir, Rajab, Shakban, Ramadhan, Syawal, Zulkaidah*, and *Zulhijah* [12].

The Hisab Urfi Islamic dating system has 30-year cycle. It contains of 11 leap years and 19 non-leap years. For leap years, a year consists of 355 days and the additional day is in the 12th month. Leap years are assigned by dividing the year number by 30 and taking the remainder. Years with a remainder of 2, 5, 7, 10, 13, 15 (or 16), 18, 21, 24, 26, or 29 are leap years [2]. In addition, there is a correction approximately every about 2347-solar years between Hisab Urfi Islamic and the Gregorian Calendar because the Hisab Urfi Islamic Calendar is one day faster than the Gregorian Calendar.

7 A 120-Year Correction

Although the Javanese Lunar Calendar has the same first date as the Hisab Urfi Islamic Calendar, the year numbering is different. In addition, they are not always synchronised. For 30 years, Javanese Lunar Calendar consists of 10,631.25 days and Hijriyah Calendar consists of 10,631 days. For 60 years, Javanese Lunar Calendar consists of 21,262.5 days and Hijriyah Calendar consists of 21,262 days. For 120 years, Javanese Lunar Calendar consists of 42,525 days and Hijriyah Calendar consists of 42,524 days. Thus, there is an additional correction every 120 years.

7.1 Date Conversion from the Solar to Javanese Calendar

To convert dates in the Western Calendar (Julian Calendar) to the Javanese Lunar Calendar, we calculate the difference in days between the date we want to convert and a reference date, ΔH. The first date is the start of the Hijriyah Calendar, i.e. Muharram 1st 1 H or July 15th or 16th 622 CE (we choose July 16th). This simplifies the conversion because the Hijriyah Calendar is related to the Javanese Lunar Calendar. The second reference date is July 16th, 1368 CE, being in the year that the ancestors of Ciptagelar civilization first saw the full Moon phase which falls on 14th each month. The date we want to convert can be located between the two reference dates or outside of both. We can choose the date from three references, i.e. Espenak, National Aeronautics and Space Administration (NASA), and Stellarium application. Full Moon dates from Stellarium will be the date where the moon has the brightest apparent magnitude. Details of this conversion are in Diagram 1.

Diagram 1. The Calculation Scheme

7.1.1 Date Conversion with the Hisab Urfi Approach

To conduct the conversion, we have to calculate ΔH at first, then treat it in the framework of Hisab Urfi Islmaic dating system. The details are as follows:

-The date we want to convert: July 1st, 1368 CE

-Visual apparent magnitude of the Moon: -12.40

Number of days between July 16th 622 CE and July 1st, 1368 CE in the Solar Calendar, ΔH:

$$\Delta H = (1368\text{ CE}-622\text{ CE}) - (\Delta\text{ days between July 1}^{st}\text{, 1368 CE and July 16}^{th}\text{, 1368 CE})$$

$$= 746\text{ years}-15\text{ days} = (746 \times 365.25)\text{ days}-15\text{ days} = 272{,}461.5\text{ days} \quad (1)$$

Number of days between July 16th, 622 CE and July 1st, 1368 CE in Hijriyah Calendar, *A*

$$A = (272{,}461.5\text{ days}) / (10{,}631\text{ days}) = 25\text{ cycles and }6{,}686.5\text{ days} \quad (2)$$

To calculate number of years in the Hijriyah Calendar *B*,

$$B = 25\text{ cycles} \times 30\text{ years/cycle} = 750\text{ H} \quad (3)$$

To calculate the month and date in the Hijriyah Calendar *C*,

$$C = 6.686{,}5{:}10.631/30 = 18\text{ H and }307.9\text{ days} \approx 18\text{ H and }308\text{ days} \quad (4)$$

$$308\text{ days} = 10\text{ months and }13\text{ days} \quad (5)$$

From *B* and *C*, to calculate the year number *D*,

$$D = 750 + 18 = 768\text{ H} \quad (6)$$

From *C* to calculate the month *E*,

$$E = 10 + 1 = 11 = \text{Month of } \textit{Dzulqo'dah} \quad (7)$$

To calculate year number in the Javanese Calendar *F*,

$$F = D + 512 = 1280\text{ Java} \quad (8)$$

Thus, July 1st, 1368 CE corresponds to *Dzulqo'da*h, 13th, 768 H and *Sela,* 13th, 1280 Java. Then we calculate the conversion for all months in 1368 CE and compare it with the two other references, i.e. Espenak and NASA. The results are in Table 2 below.

Table 2. Date Conversion from Julian Calendar to Hijriyah Calendar and Javanese Calendar for 1368 CE with *Hisab Urfi* Approach

Julian Calendar (1368 CE)				Hijriyah Calendar			Javanese Calendar		
Literature	Month	Date	UTC +07	Year	Month	Date	Year	Month	Date
Espenak	Jan	6	03:26	768	Jumadil Awal	13	1280	Jumadil Awal	13
NASA		-	-	-	-	-	-	-	-

Julian Calendar (1368 CE)				**Hijriyah Calendar**			**Javanese Calendar**		
Literature	**Month**	**Date**	**UTC +07**	**Year**	**Month**	**Date**	**Year**	**Month**	**Date**
Stellarium		6	23:00	768	Jumadil Awal	13	1280	Jumadil Awal	13
Espenak		4	22:39	768	Jumadil Akhir	12	1280	Jumadilakir	12
NASA	*Feb*	-	-	-	-	-	-	-	-
Stellarium		5	23:00	768	Jumadil Akhir	13	1280	Jumadilakir	13
Espenak		5	15:22	768	Rajab	13	1280	Rejeb	13
NASA	Mar	-	-	-	-	-	-	-	-
Stellarium		5	23:00	768	Rajab	13	1280	Rejeb	13
Espenak		4	04:54	768	Syakban	13	1280	Ruwah	13
NASA	Apr	-	-	-	-	-	-	-	-
Stellarium		4	23:00	768	Syakban	13	1280	Ruwah	13
Espenak		3	15:36	768	Ramadhan	13	1280	Pasa	13
NASA	May	-	-	-	-	-	-	-	-
Stellarium		3	23:00	768	Ramadhan	13	1280	Pasa	13
Espenak		2	00:18	768	Syawal	13	1280	Sawal	13
NASA	Jun	-	-	-	-	-	-	-	-
Stellarium		2	23:00	768	Syawal	13	1280	Sawal	13
Espenak		1	07:49	768	Dzulqo'dah	13	1280	Sela	13
NASA	Jul (1)	-	-	-	-	-	-	-	-
Stellarium		1	23:00	768	Dzulqo'dah	13	1280	Sela	13
Espenak		30	15:01	768	Dzulhijah	12	1280	Besar	12
NASA	*Jul (2)*	-	-	-	-	-	-	-	-
Stellarium		31	23:00	768	Dzulhijah	13	1280	Besar	13
Espenak		28	22:43	769	Muharram	12	1281	Sura	12
NASA	*Agt*	-	-	-	-	-	-	-	-
Stellarium		29	23:00	769	Muharram	13	1281	Sura	13
Espenak		27	07:54	769	Safar	12	1281	Sapar	12
NASA	*Sep*	-	-	-	-	-	-	-	-
Stellarium		28	23:00	769	Safar	13	1281	Sapar	13
Espenak	*Oct*	26	19:28	769	Rabi'ul Awal	12	1281	Mulud	12
NASA		-	-	-	-	-	-	-	-

Julian Calendar (1368 CE)				Hijriyah Calendar			Javanese Calendar		
Literature	**Month**	**Date**	**UTC +07**	**Year**	**Month**	**Date**	**Year**	**Month**	**Date**
Stellarium		27	23:00	769	Rabi'ul Awal	13	1281	Mulud	13
Espenak		25	10:01	769	Rabi'ul Akhir	12	1281	Bakda Mulud	12
NASA	*Nov*	-	-	-	-	-	-	-	-
Stellarium		26	23:00	769	Rabi'ul Akhir	13	1281	Bakda Mulud	13
Espenak		25	03:17	769	Jumadil Awal	13	1281	Jumadil Awal	13
NASA	Des	-	-	-	-	-	-	-	-
Stellarium		25	23:00	769	Jumadil Awal	13	1281	Jumadil Awal	13

Note: *Italic* means there is inequality of date between those two references.

7.1.2 Date Conversion Using A 120-Year Correction between Hijriyah and Javanese Calendar

To conduct this conversion, we use again (1) then treat it in the framework of the Javanese Calendar. The details are as follows:

Number of days between July 16th, 622 CE and July 1st, 1368 CE in Javanese Calendar *G*,

$$G = 272{,}461.5/2{,}835 = 96 \text{ cycles and } 301.5 \text{ days} \tag{9}$$

To calculate number of years in the Javanese Calendar that has passed *H*,

$$H = 96 \text{ cycles x } 8 \text{ years/cycle} = 768 \text{ Java} \tag{10}$$

Using 512 as the difference in the number of years between the Hijriyah and Javanese Calendar, the actual Javanese year number *J*,

$$J = H + 512 = 1280 \text{ Java} \tag{11}$$

To calculate the month and date in Javanese Calendar *K*,

$$K = 301.5 \text{ days} = 10 \text{ months and } 6.5 \text{ days} \approx 10 \text{ months and } 7 \text{ days} \tag{12}$$

From *K*, number of month *L*,

$$L = 10 + 1 = 11 = \textit{Sela} \tag{13}$$

With a 120-year correction *N*, we obtain,

$$N = 7 + 6 = 13 \tag{14}$$

Thus July 1st, 1368 CE corresponds to *Sela* 13th, 1280 Java. Then we calculate the conversion for all months in 1368 CE in the framework of the Javanese Calendar and compare it with the two other references, i.e. Espenak and NASA. The results are shown in the Table 3.

Table 3. Date Conversion from Julian Calendar to Javanese Calendar for 1368 CE with a 120-Year Correction

Julian Calendar (1368 CE)				**Javanese Calendar**		
Literature	**Month**	**Date**	**UTC +07**	**Year**	**Month**	**Date**
Espenak	Jan	6	3:26	1280	Jumadil Awal	13
NASA		-	-	-	-	-
Stellarium		6	23:00	1280	Jumadil Awal	13
Espenak	*Feb*	4	22:39	1280	Jumadilakir	12
NASA		-	-	-	-	-
Stellarium		5	23:00	1280	Jumadilakir	13
Espenak	Mar	5	15:22	1280	Rejeb	13
NASA		-	-	-	-	-
Stellarium		5	23:00	1280	Rejeb	13
Espenak	Apr	4	4:54	1280	Ruwah	13
NASA		-	-	-	-	-
Stellarium		4	23:00	1280	Ruwah	13
Espenak	May	3	15:36	1280	Pasa	13
NASA		-	-	-	-	-
Stellarium		3	23:00	1280	Pasa	13
Espenak	Jun	2	0:18	1280	Sawal	13
NASA		-	-	-	-	-
Stellarium		2	23:00	1280	Sawal	13
Espenak	Jul (1)	1	7:49	1280	Sela	13
NASA		-	-	-	-	-
Stellarium		1	23:00	1280	Sela	13
Espenak	*Jul (2)*	30	15:01	1280	Besar	12
NASA		-	-	-	-	-
Stellarium		31	23:00	1280	Besar	13
Espenak	*Agt*	28	22:43	1281	Sura	12
NASA		-	-	-	-	-
Stellarium		29	23:00	1281	Sura	13
Espenak	*Sep*	27	7:54	1281	Sapar	12
NASA		-	-	-	-	-
Stellarium		28	23:00	1281	Sapar	13

Julian Calendar (1368 CE)				Javanese Calendar		
Literature	**Month**	**Date**	**UTC +07**	**Year**	**Month**	**Date**
Espenak		26	19:28	1281	Mulud	12
NASA	*Oct*	-	-	-	-	-
Stellarium		27	23:00	1281	Mulud	13
Espenak		25	10:01	1281	Bakda Mulud	12
NASA	*Nov*	-	-	-	-	-
Stellarium		26	23:00	1281	Bakda Mulud	13
Espenak		25	3:17	1281	Jumadilawal	13
NASA	Des	-	-	-	-	-
Stellarium		25	23:00	1281	Jumadilawal	13

Note: *Italic* means there is inequality of date between those two references.

7.1.3 Date Conversion from Solar to Javanese Calendar for 2018 CE

To calculate this conversion, we take into account that the Western Calendar transitioned from the Julian Calendar to the Gregorian Calendar on October 5th, 1582 CE. Using the same conversion as the previous section, we get for the last day of the Julian Calendar,

October 4th, 1582 CE = *Ramadhan* 16th, 990 H = *Pasa* 16th, 1502 Java.

We know that the next day is *Pasa* 17th, 1502 Java, but in the Gregorian Calendar the next day is October 15th, 1582 CE. For this reason, we choose new reference dates for the conversion, indicated in Diagram 2.

Diagram 2. The Calculation Scheme

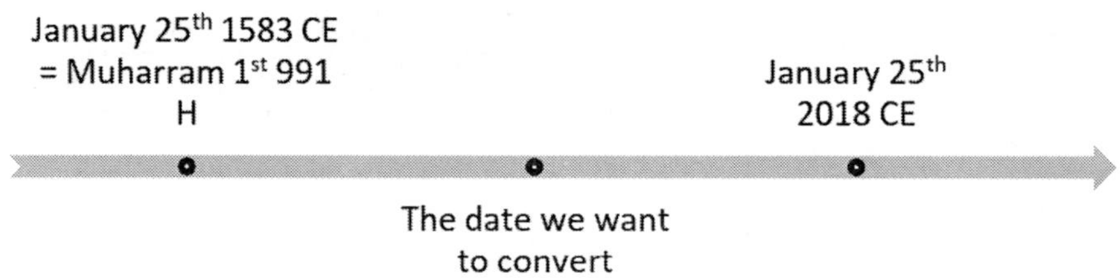

As we did for the previous conversion, we calculate for all months in 2018 CE in the framework of the Javanese Calendar using a 120-year correction and compare it with the two other references. We can see the result in the Table 4 below.

Table 4. Date Conversion from Gregorian Calendar to Hijriyah Calendar and Javanese Calendar in 2018 CE

Gregorian Calendar (2018 CE)				Hijriyah Calendar			Javanese Calendar		
Literature	**Month**	**Date**	**UTC +07**	**Year**	**Month**	**Date**	**Year**	**Month**	**Date**
Espenak	Jan (1)	2	9:24	1439	Rabi'ul Akhir	13	1951	Bakda Mulud	12

Gregorian Calendar (2018 CE)				Hijriyah Calendar			Javanese Calendar		
Literature	**Month**	**Date**	**UTC +07**	**Year**	**Month**	**Date**	**Year**	**Month**	**Date**
NASA		2	9:24	1439	Rabi'ul Akhir	13	1951	Bakda Mulud	12
Stellarium		2	23:00	1439	Rabi'ul Akhir	13	1951	Bakda Mulud	12
Espenak	Jan (2)	31	20:27	1439	Jumadil Awal	13	1951	Jumadil Awal	12
NASA		31	20:27	1439	Jumadil Awal	13	1951	Jumadil Awal	12
Stellarium		31	23:00	1439	Jumadil Awal	13	1951	Jumadil Awal	12
Espenak	*Mar (1)*	2	7:51	1439	Jumadil Akhir	14	1951	Jumadilakir	13
NASA		2	7:51	1439	Jumadil Akhir	14	1951	Jumadilakir	13
Stellarium		1	23:00	1439	Jumadil Akhir	13	1951	Jumadilakir	12
Espenak	Mar (2)	31	19:37	1439	Rajab	13	1951	Rejeb	12
NASA		31	19:37	1439	Rajab	13	1951	Rejeb	12
Stellarium		31	23:00	1439	Rajab	13	1951	Rejeb	12
Espenak	Apr	30	7:58	1439	Syakban	13	1951	Ruwah	12
NASA		30	7:58	1439	Syakban	13	1951	Ruwah	12
Stellarium		30	23:00	1439	Syakban	13	1951	Ruwah	12
Espenak	May	29	21:20	1439	Ramadhan	13	1951	Pasa	12
NASA		29	21:20	1439	Ramadhan	13	1951	Pasa	12
Stellarium		29	23:00	1439	Ramadhan	13	1951	Pasa	12
Espenak	Jun	28	11:53	1439	Syawal	13	1951	Sawal	12
NASA		28	11:53	1439	Syawal	13	1951	Sawal	12
Stellarium		28	23:00	1439	Syawal	13	1951	Sawal	12
Espenak	*Jul*	28	3:21	1439	Dzulqo'dah	14	1951	Sela	13
NASA		28	3:21	1439	Dzulqo'dah	14	1951	Sela	13
Stellarium		27	23:00	1439	Dzulqo'dah	13	1951	Sela	12
Espenak	Agt	26	18:56	1439	Dzulhijah	13	1951	Besar	12
NASA		26	18:56	1439	Dzulhijah	13	1951	Besar	12
Stellarium		26	23:00	1439	Dzulhijah	13	1951	Besar	12
Espenak	Sep	25	9:53	1440	Muharram	13	1952	Sura	12
NASA		25	9:53	1440	Muharram	13	1952	Sura	12
Stellarium		25	23:00	1440	Muharram	13	1952	Sura	12
Espenak	*Oct*	24	23:45	1440	Safar	12	1952	Sapar	11

Gregorian Calendar (2018 CE)				**Hijriyah Calendar**			**Javanese Calendar**		
Literature	**Month**	**Date**	**UTC +07**	**Year**	**Month**	**Date**	**Year**	**Month**	**Date**
NASA		24	23:45	1440	Safar	12	1952	Sapar	11
Stellarium		25	23:00	1440	Safar	13	1952	Sapar	12
Espenak		23	12:39	1440	Rabi'ul Awal	13	1952	Mulud	12
NASA	Nov	23	12:39	1440	Rabi'ul Awal	13	1952	Mulud	12
Stellarium		23	23:00	1440	Rabi'ul Awal	13	1952	Mulud	12
Espenak		23	0:49	1440	Rabi'ul Akhir	13	1952	Bakda Mulud	12
NASA	Des	23	0:49	1440	Rabi'ul Akhir	13	1952	Bakda Mulud	12
Stellarium		23	23:00	1440	Rabi'ul Akhir	13	1952	Bakda Mulud	12

Note: *Italic* means there is inequality of date between those two references.

We find from this table that every date in the Gregorian Calendar that we convert into the Hijriyah and the Javanese Lunar Calendar, there is no same date between the Hijriyah and the Javanese Lunar Calendar even though we have applied the 120-year correction. From Table 4, we find that the Javanese Lunar Calendar is one day slower than the Hijriyah calendar. This invalidity happens since 2010 CE, the cause of which is still unknown.

8 Conclusions

We conclude that the date of the *Malam Opat Belasan* ceremony which is always held on the 13th to the night of 14th in every month is correct if the date of the full Moon phase that we choose in Stellarium follows the definition of full Moon phase in the Hijriyah Calendar. Specifically, the Hijriyah definition of the full moon indicates that the full Moon phase occurs on the 13rd, 14th, and 15th day of each lunar month.

However, we also find that the 120-year correction between the Hijriyah and Javanese Lunar Calendar is invalid since 2010 CE. The Javanese Lunar Calendar is one day slower than Hijriyah calendar. The cause of this is still unknown.

References

1. S. D. Wiramihardja, *Ethnoastronomy: The Sundanese of West Jawa and their Relation to Ethnoastronomy*, **68**, 25-27 (2013)
2. M. Raharto, *Sistem Kalendar: Pengantar Studi Hubungan Kalendar dan Fenomena Astronomi Sistem Bumi, Bulan dan Matahari* (2005)
3. I. Hakim, *Kalender Jawa dan Sunda* (2014)
4. G. Ammarel, L. Tsing, *Cultural Production of Skylore in Indonesia* Research Get 2207-2214 doi: 10.1007/978-1-4614-6141-8_236 (2015)
5. G. Ammarel, *Sky Calendars of The Indo-Malay Archipelago: Regional Diversity/Local Knowledge* 85-104 (2015)
6. S. T. T. Raffles, *The History of Java*, **2**, (3)

7. C. M. D. Silva, *The Spring Full Moon*, 475-478 (2004)
8. M. Zeilik, *The Ethnoastronomy of The Historic Pueblos, II: Moon Watching*, **10**, 1-22 (1986)
9. M. Vahia, *Ethno Astronomy in India. Ethnoastronomy Awakened in Mae Taeng District-Stampeding at Global Arena* (2017)
10. P. Dahedar, M. Vahia, *Fieldwork Techniques in Ethnoastronomy. Ethnoastronomy Awakened in Mae Taeng District-Stampeding at Global Arena* (2017)
11. W. Orchiston, *Studying The History of Indonesian Astronomy: Future Prospects and Possibilities*, **20**, (2), 145-154 (2017)
12. Jayusman *Wacana Takwim Urfi dalam Penanggalan Islam*, **7**, 18-30 (2009)
13. G. H. Iskandar, *A Notes From Kasepuhan Ciptagelar: Revisiting Pranatamangsa*, 1-14 (2016)
14. R. M. R. B. Setiawan, *Hitungan Kalender Jawa Menurut Sultan Agung*
15. https://ciptagelar.info/events/opat-belasan/ accessed on July 24th 2018 at 02:10 p.m.
16. http://www.pikiran-rakyat.com/jawa-barat/2017/03/24/tradisi-ciptagelar-di-sukabumi-mampu-pertahankan-kampung-adat-397144 accessed on July 24th 2018 at 03:07 p.m
17. https://www.kompasiana.com/muhiqaz/54f84d79a33311d25d8b4a0e/keunikan-perkampungan-di-lereng-gunung-halimun accessed on July 25th 2018 at 07:02 p.m
18. https://gstff.wordpress.com/2016/02/08/menuju-gunung/amp/ accessed on July 27th 2018 at 05:50 p.m.

Galactic warp from the kinematics of OB stars

Rizky Maulana Nurhidayat[1,*], *Mochamad Ikbal* Arifyanto[2,**], and *Lucky* Puspitarini[2]

[1]Astronomy Study Program, Faculty of Mathematics and Natural Sciences, Institut Teknologi Bandung, Indonesia

[2]Astronomy Research Division, Faculty of Mathematics and Natural Sciences, Institut Teknologi Bandung, Indonesia

Abstract. The warp in Milky Way seen through the young stellar disk shows the complex structures. Its spatial distribution reveals the asymmetry in the vertical distance to the galactic midplane, whereas the kinematics shows the oscillating motion. We analyze the warp using about 25,000 OB stars from Gaia Data Release 2, with Galactocentric azimuth (ϕ) range from 135° to 225°. We detect the warp from two distinctive regions. First region (warp down) lies on $100° < l < 150°$ and second regions (warp up) lies on $150° < l < 220°$. The vertical position towards the Galactic midplane (Z) from all samples shows the line-of-nodes of the warp is on $\phi = 174° \pm 2°$. The maximum deviation in warp down region reaches 0.1 kpc and 0.2 kpc in warp up region. The vertical velocity (V_Z) from both regions are dominated by negative V_Z. From the kinematics distribution, it can be concluded that there is an oscillati1ng motion. Considering with the previous study, the oscillating motion was caused by the gravitational interaction with the Galaxy satellites. From this, we can constrain the origin of the warp is developing from the tidal interaction between Milky Way and its satellites.

1 Introduction

A common feature in a spiral galaxy is a warped disk. Analyzing edge-on spiral galaxies from Sloan Digital Sky Survey and NASA Extragalactic Database, there are 236 from 325 galaxies that develop warp, with 165 galaxies have S-shaped warp and 71 galaxies have U-shaped warp [1]. Warp also develops in the nearest galaxy, M31. The integral-sign warp starts at $R \simeq 30$ kpc from the galaxy center of M31 with a deviation of more than 3 kpc to the galaxy midplane [2].

The study about warp in Milky Way comes from the analysis of the spatial distribution of HI gas [3] and the young-old stellar population [4]. They found that the Milky Way develops aligned warp both in the HI gas and stellar distribution. The second quadrant of Galactocentric azimuth (ϕ) is dominated by gas and stars located above the Galaxy midplane. The opposite features are shown in the third quadrant. The deviation of the warp is dependent on the tracers' age [4]. The warp from the young stellar population (OB stars) deviates of 0.23 kpc and 0.19 kpc on the warp up and warp down region, respectively. The warp developing in the old stellar population (RGB stars) reaches up to 0.97 kpc at the warp up region and

*e-mail: rizman.nana@students.itb.ac.id
**e-mail: ikbal@as.itb.ac.id

1.22 kpc at the warp down region [4]. From another young stellar tracer like Cepheid stars, the warp starting develops at 9.26 kpc with line-of-nodes (LoN) of 17.4° [5]. The kinematics distribution of Upper Main Sequence (UMS) that is called OB stars in this paper and Red Giant Branch (RGB) star in Gaia Data Release 2 (DR2) shows the increasing vertical velocity in the anti-center direction [6].

The origin of the warp is still debatable. It can be a product of a misaligned inner and outer halos in the dynamics of the galaxy [7], the misaligned of the angular momentum axis with the disk spin axis of the late infall [8], tidal interaction between galaxy satellites and its host [9, 10] and the interaction between galaxy magnetic fields with the intergalactic magnetic fields [11]. We collect the phase-space motion of OB stars from Gaia DR2 [12] cross-matched with LAMOST DR4 [13], RAVE DR5 [14], GALAH DR2 [15], 2nd. Catalog of Radial Velocities with Astrometric Data (hereafter CRVAD) [16], and Pulkovo Radial Velocities for 35493 HIP Stars (hereafter PRV) [17] to analyze the Milky Way warp and constrain its origin.

2 Data and Methods

To select OB stars from Gaia DR2, we apply initial criteria as mentioned in [18], shown in equation (1) and (2).

$$M_G = G + 5\log\varpi - A_G - 10 < 2 \tag{1}$$

$$(G_{BP} - G_{RP})_0 = (G_{BP} - G_{RP}) - E(G_{BP} - G_{RP}) < 0, \tag{2}$$

where ϖ is the parallax of the stars in mas, G, G_{BP}, G_{RP}, $E(G_{BP} - G_{RP})$, and A_G are photometric in G-band (330–1050 nm), G_{BP} (330–680 nm), G_{RP} (630–1050 nm), color excess, and extinction were provided by Gaia DR2, respectively.

The next step is excluding the contamination of giants or red clump stars in OB star candidates we get from equation (1) and (2), with applying equation (3) and (4).

$$J - H < 0.14(G - K_S) + 0.02 \tag{3}$$

$$J - K_S < 0.23(G - K_S), \tag{4}$$

where J, H, and K_S are the photometric in JHK_S-bands provided by 2MASS. The distance for each OB stars is calculated with equation (5) that is valid for $\varpi > 0$ and $\varpi_{err}/\varpi < 0.2$ as mentioned in [19], where ϖ_{err} is the uncertainty of parallax in mas.

$$D = \frac{1}{\varpi}, \tag{5}$$

where D is in kpc. To ensure our OB samples, we exclude the stars which have vertical distance from galactic midplane (|Z|) of more than 2 kpc. Finally, we find the radial velocity from OB stars that not provided in Gaia DR2 by cross-matching with LAMOST DR4, RAVE DR5, GALAH DR2, CRVAD, and PRV. We get phase-space coordinates of 24,958 OB stars.

We use a galaxy model with the distance of Sun-Galactic Center $X_\odot = -8.34$ kpc [20] and the distance of Sun-Galactic midplane $Z_\odot = 0.027$ kpc [21]. The coordinate transformation, from Galactic coordinates to cartesian Galactocentric coordinates, is shown in equation (6), (7), and (8).

$$X = D\cos b\cos l - X_\odot \tag{6}$$

$$Y = D\cos b\sin l \tag{7}$$

$$Z = D\sin b + Z_\odot, \tag{8}$$

where l and b are the Galactic coordinates of each star, X, Y, and Z is the position concerning the Galactic Center (GC), adopting the right-handed system. To transform in cylindrical Galactocentric, we apply equation (9) and (10).

$$R = \sqrt{X^2 + Y^2} \tag{9}$$

$$\phi = \arcsin\left(\frac{D \sin l}{R}\right). \tag{10}$$

We analyze the distribution of Milky Way warp from the XY projection with the color-coded of Z. Then, we find LoN of the warp from our samples with determining the Z as a function of ϕ using the transformation in equation (10), with R is the Galactocentric radius and ϕ that starting from positive of X-axis increases in the opposite direction of the galactic rotation. We bin the data for every 500 stars, with no overlapping. To analyze the kinematics distribution of the warp from our samples, we measure the vertical velocity (V_Z) as a function of ϕ, with no overlapping 150 stars for each bin.

3 Results and Analysis

The spatial distribution of our OB stars in the XY projection is shown in Figure 1. Our OB stars is distributed in about $135° < \phi < 225°$. Figure 1 shows the complex distribution, in line with the previous study [4]. The color-coded in Figure 1 shows the mean Z for each bin with a size of 0.1 kpc × 0.1 kpc. The solid gray line indicates the Galactocentric radius of 8, 10, 12, and 14 kpc with a red dot is the position of the Sun. We detect the warp developed in the blue and red regions. The blue region called warp up region, lies on Galactic longitude $100° < l < 150°$, and the red region called warp down region, lies on $150° < l < 220°$.

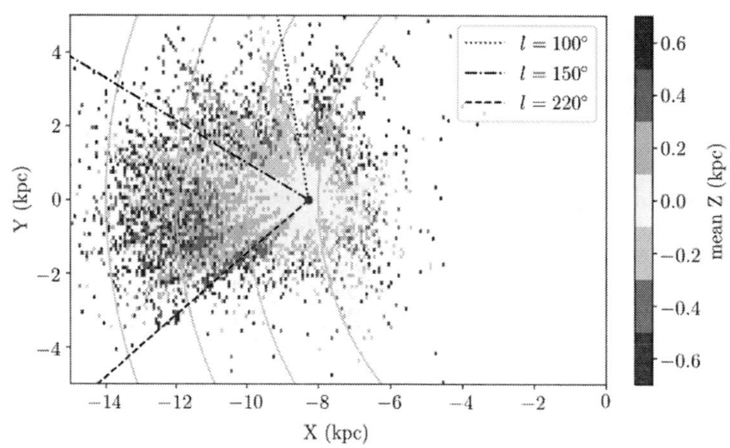

Figure 1. XY projection of OB stars in about $135° < \phi < 225°$. Four gray solid lines are $R = 8, 10, 12$, and 14 kpc with mean Z as a color-code. Sun is presented by a red dot. Warp down and warp up region are selected in $100° < l < 150°$ and $150° < l < 220°$, respectively.

In analyzing LoN of the warp from our OB stars, we measure the mean of Z as a function of ϕ, shown in Figure 2. The error bar for each data point comes from the standard deviation in each bin. We can determine LoN lies at $\phi = 174° \pm 2°$. The uncertainty of LoN is calculated by the mean ϕ of two nearest data point to $Z = 0$ kpc. From this result, the Sun is not located on LoN, in line with the previous study [22]. The maximum deviation below the galactic midplane reaches about 0.1 kpc, at $\phi = 170°$ and the deviation above the galactic midplane reaches 0.2 kpc at $\phi = 190°$. The asymmetry of Milky Way warp also found in the previous study [4, 23], with different representations. If we assume the mean Z in Figure 2 can be approximated with sinusoidal function, several data points in $177° < \phi < 182°$ will be out of the function. We guess that it is caused by the presence of the North Near (NN) overdensity [24]. We get more stars in NN lies on $177° < \phi < 182°$ and one data point at $\phi = 187°$, but less stars in other region of NN.

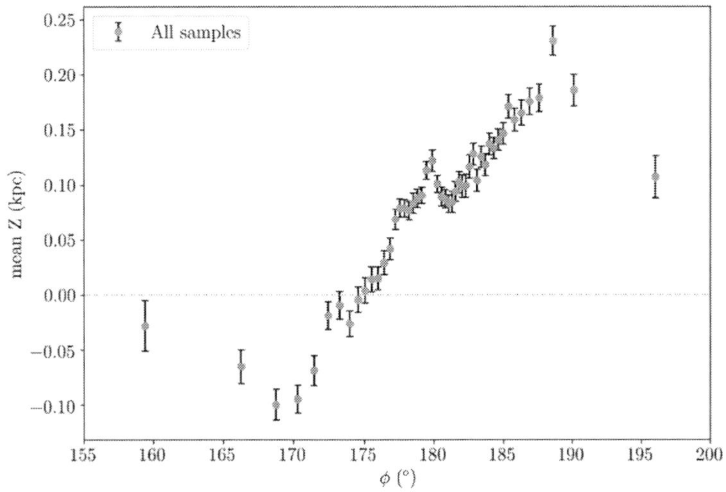

Figure 2. Mean Z as a function of ϕ for all samples. Each data point is consisted of 500 stars, with no overlapping. The gray dashed line is the position of galactic midplane.

The kinematics distribution of the warp is shown in Figure 3. We analyze two regions found in Figure 1, warp down and warp up region are represented by red and blue data point, respectively. The binning method is similar to the previous binning applied in Figure 2, with the number of stars for each bin is 150. The kinematics distribution of the warp down region, ($Z < 0$), is dominated by the negative V_Z. We get a similar distribution in the warp up region, that has $Z > 0$. It can be expected that both of the regions are not static. It is like a part of oscillation. Because of this oscillation, the LoN of Milky Way warp will be shifted. Comparing with the previous work [25], we guess that the oscillation we found is just a small part of vertical damped oscillation in [25]. The damped oscillation is expected to come from the interaction between dwarf Sagittarius with Milky Way. So, we constrain the origin of the warp we detected is coming from the dwarf Sagittarius interaction in orbiting Milky Way.

4 Conclusion

The Milky Way warp from OB stars reveals a complex spatial distribution. In the kinematics distribution, we can conclude that the warp is not a static feature. The warp down and warp up region we found are dominated by negative V_Z. It shows the vertical oscillation motion of the stars, just in a small region. Considering the study of the perturbed Milky Way disk, our result is in line with the previous study [25]. From these oscillations, the origin of the Milky Way warp can be constrained. It could be a product of the interaction of the galactic satellites

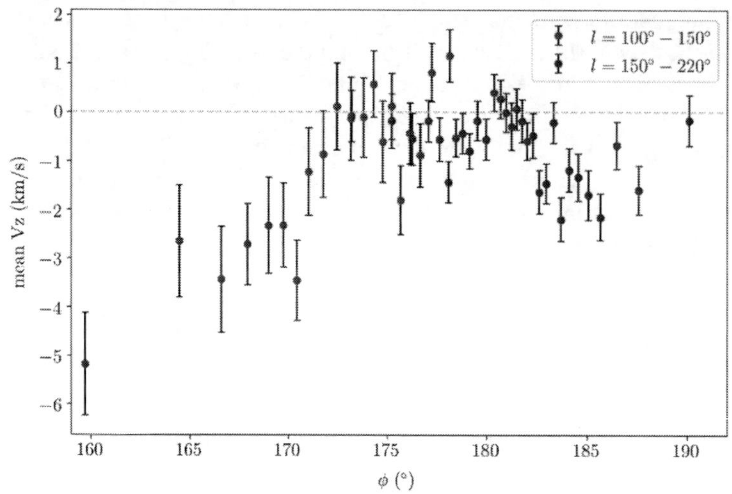

Figure 3. Mean V_Z as a function of ϕ for warp down and warp up region. Warp down and warp up region are presented by red dots and blue dots, respectively. Each data point is consisted of 150 stars, with no overlapping.

with Milky Way. In this case, dwarf Sagittarius that has a polar orbit is expected as a culprit. A more detailed study is needed to ensure the dynamics of the Milky Way warp in the past by analyzing the spatial and kinematics from the different stellar populations and building a model of the warp.

5 Acknowledgement

The authors gratefully acknowledge a research grant from the Program of Research, Community Service, and Innovation of the Institut Teknologi Bandung (P3MI-ITB). This work has made use of data from the European Space Agency (ESA) mission *Gaia* (https://www.cosmos.esa.int/gaia), Guoshoujing Telescope (the Large Sky Area Multi-Object Fiber Spectroscopic Telescope LAMOST) survey, Radial Velocity Experiment (RAVE) survey, The Galactic Archaeology with HERMES (GALAH) survey, VizieR Online Data Catalog: 2nd Cat. of Radial Velocities with Astrometric Data, and VizieR Online Data Catalog: Pulkovo radial velocities for 35493 HIP stars.

References

[1] H.B. Ann, J-C. Park, New Astron. **11**, 293 (2006)
[2] K. Newton, D.T. Emerson, MNRAS **181**, 573 (1977)
[3] E.S. Levine, L. Blitz, C. Heiles, ApJ **643**, 881 (2006)
[4] M. Romero-Gómez, C. Mateu, L. Aguilar, F. Figueras, A. Castro-Ginard, A&A **627**, A150 (2019)
[5] X. Chen, S. Wang, L. Deng, R. de Grijs, C. Liu, H. Tian, Nature Astronomy **3**, 320 (2019)
[6] E. Poggio, R. Drimmel, M.G. Lattanzi, R.L. Smart, A. Spagna, et al., MNRAS **481**, L21 (2018)
[7] J. Dubinsky, D. Chakrabarty, ApJ **703**, 2068 (2009)

[8] J.A. Sellwood, TD Oswalt & G. Gilmore **923**, (2013)
[9] F.A. Gómez, I. Minchev, B.W. O'Shea, T.C. Beers, J.S. Bullock, C.W. Purcell, MNRAS **429**, 159 (2012)
[10] J.H. Kim, S. Peirani, S. Kim, H.B. Ann, S. An, S. Yoon, ApJ **789**, 90 (2014)
[11] E. Battaner, J. Jimenez-Vicente, arXiv preprint astro-ph/9802020 (1998)
[12] Gaia Collaboration, T. Prusti, J.H.J. De Bruijne, A.G.A. Brown, A. Vallenari, et al., A&A **595**, A1 (2018a)
[13] A.L. Luo, Y.H. Zhao, G. Zhao, et al., Vizier Online Data Catalog (2019)
[14] A. Kunder, G. Kordopatis, M. Steinmetz, T. Zwitter, et al., AJ **153**, 75 (2016)
[15] S. Buder, M. Asplund, L. Duong, J. Kos, et al., MNRAS **478**, 4513 (2018)
[16] N.V. Kharchenko, R.D. Scholz, A.E. Piskunov, S. Roeser, E. Schilbach, Vizier Online Data Catalog (2007)
[17] G.A. Gontcharov, Vizier Online Data Catalog (2006)
[18] Gaia Collaboration, D. Katz, T. Antoja, M. Romero-Gómez, et al., A&A **616**, A11 (2018b)
[19] C.A.L. Bailer-Jones, PASP **127**, 994 (2015)
[20] M.J. Reid, K.M. Menten, A. Brunthaler, et al., ApJ **783**, 130 (2014)
[21] B. Chen, C. Stoughton, J. A. Smith, et al., ApJ **553**, 184 (2001)
[22] C. Li, G. Zhao, Y. Jia, S. Liao, C. Yang, Q. Wang, ApJ **871**, 208 (2019)
[23] I. Yusifov, *The Magnetized Interstellar Medium*: Proc. Conference held in Antalya, Turkey (eds Uyaniker, B., Reich, W. & Wielebinski, R.) 165-169 (Copernicus GmbH, 2004).
[24] Y. Xu, H.J. Newberg, J.L. Carlin, C. Liu, L. Deng, J. Li, R. Schönrich, B. Yanny, ApJ **801**, 105 (2015)
[25] T. Antoja, A. Helmi, M. Romero-Gómez, D. Katz, C. Babusiaux, R. Drimmel, D.W. Evans, F. Figueras, E. Poggio, C. Reylé, A.C. Robin, G. Seabroke, C. Soubiran, Nature **561**, 360 (2018)

Numerical solutions to Giovannini's QCD parton branching processes

Z. Ong[1,*], *P.* Agarwal[1], *H.W.* Ang[1], *E.* Wang[1], *A.H.* Chan[1], and *C.H.* Oh[1]

[1]Department of Physics, National University of Singapore. 2 Science Drive 3, Singapore 117551.

Abstract. In 1979, Giovannini [1] formulated multiparticle production as a manifestation of four QCD processes: $A : g \to gg$ (gluon bremsstrahlung), $\widetilde{A} : q \to qg$ (quark bremsstrahlung), $B : g \to \overline{q}q$ (quark pair creation) and $C : g \to ggg$ (four-gluon vertex). At present, only approximate solutions exist, such as the Generalized Multiplicity Distribution (GMD) which excludes processes B and C [2, 3]. These approximate solutions have been very successful in describing charged particle multiplicities at lower energies, at least until the appearance of a shoulder-like structure at $\sqrt{s} \approx 200$ GeV and 900 GeV first reported by the UA5 collaboration [4], also known as "KNO-scaling violation". Here, we extend the work of Sakai [5] and attempt a numerical solution that incorporates all four QCD processes, and see if the addition of processes B and C can explain the shoulder-like structure.

Introduction

Giovannini's formulation of multiparticle production (as described in the abstract) can be cast into a differential-difference equation. For m quarks and n gluons at QCD evolution parameter t, we have:

$$\begin{aligned}\frac{\partial P_{m,n}}{\partial t} = &- (An + \widetilde{A}m + Bn + Cn)P_{m,n} \\ &+ A(n-1)P_{m,n-1} \\ &+ \widetilde{A}mP_{m,n-1} \\ &+ B(n+1)P_{m-2,n+1} \\ &+ C(n-2)P_{m,n-2}\end{aligned} \tag{1}$$

where A, $\widetilde{A}$, B and C are the probabilities of the processes taking place. To investigate how KNO-scaling violation might arise from these four fundamental processes, we will integrate eq. (1) numerically using the 4th-order Runge-Kutta method (RK4) and study how varying A, $\widetilde{A}$, B and C modifies the final distribution, observing for the process(es) that might be responsible for producing the shoulder-like structure.

*e-mail: ongzongjin@u.nus.edu

Procedure

First, we set up a 2-dimensional probability distribution $P_{2D}[m][n]$, where m = number of quarks and n = number of gluons. For an initial condition of 2 quarks and 0 gluons (as an example), we set $P_{2D}[2][0] = 1$ and all other $P_{2D}[m][n] = 0$. The entire array is then integrated element-wise using RK4 from $t = 0.0$ to 3.0 (the upper limit is arbitrarily set), with step size $h = 0.01$.

$P_{2D}[m][n]$ is then condensed into a parton distribution $P_{parton}(n)$, where both quarks and gluons are categorised together. $P_{parton}(n)$ is truncated at $n = MX$ (identified as the "kinematical bound" in [5]), which is set at $MX = 100$ in this work to reflect finite branching. The truncated $P_{parton}(n)$ is then normalised to 1.

Next, we convert the parton distribution to an all-hadron distribution $P_h(n)$. Each parton hadronises into k hadrons according to a Poisson distribution (with average value $\lambda = 2.5$ and maximum value $k_{max} = 7$, following [5]):

$$P(k) = \frac{\frac{\exp(-\lambda)\cdot\lambda^k}{k!}}{\sum_{k=1}^{7} \frac{\exp(-\lambda)\cdot\lambda^k}{k!}} \tag{2}$$

The all-hadron distribution $P_h(n)$ is then converted into a charged-hadron distribution $P_c(n_c)$. As before, we adopt the model used in [5]: pairs of $(+,-)$ or $(0,0)$ occur in equal probability, and we assume that the initial state contains an even number of charged particles. This constraints the final state to contain an even number of charged particles due charge conservation. If the final state has $(2n + 1)$ hadrons, the final odd hadron must then be neutral. The charged-hadron distribution is then computed from the all-hadron distribution using combinatorics:

$$P_c(2n_c) = \sum_{n \geq n_c} \frac{n!}{n_c!(n-n_c)!2^l} [P_h(2n) + P_h(2n+1)] \tag{3}$$

Plots

In each of the plots that follow, we vary a selected parameter against a base configuration: $A = 0.2$, $\widetilde{A} = 0.2$, $B = 0$, $C = 0$, $M = 2$ and $N = 0$. Each figure shows a selected parameter taking three different values (one low, one intermediate and one high, labelled within the plot), with the other parameters taking the default values from the base configuration.

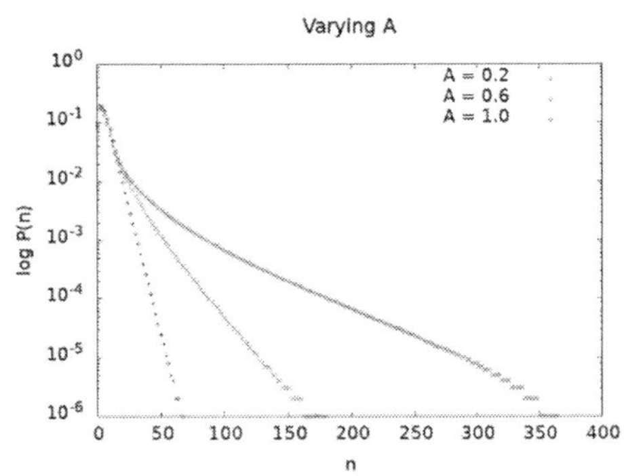

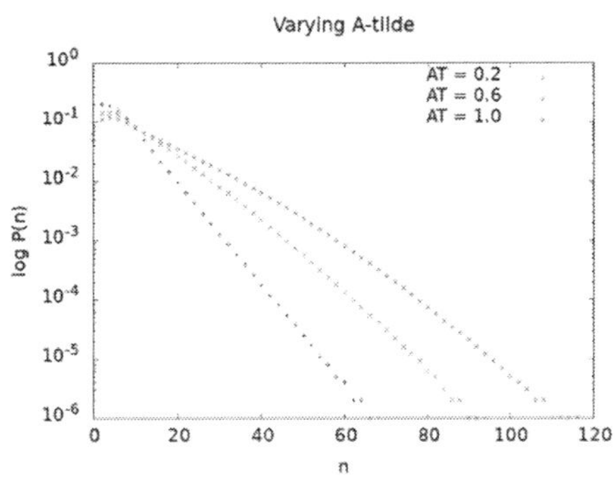

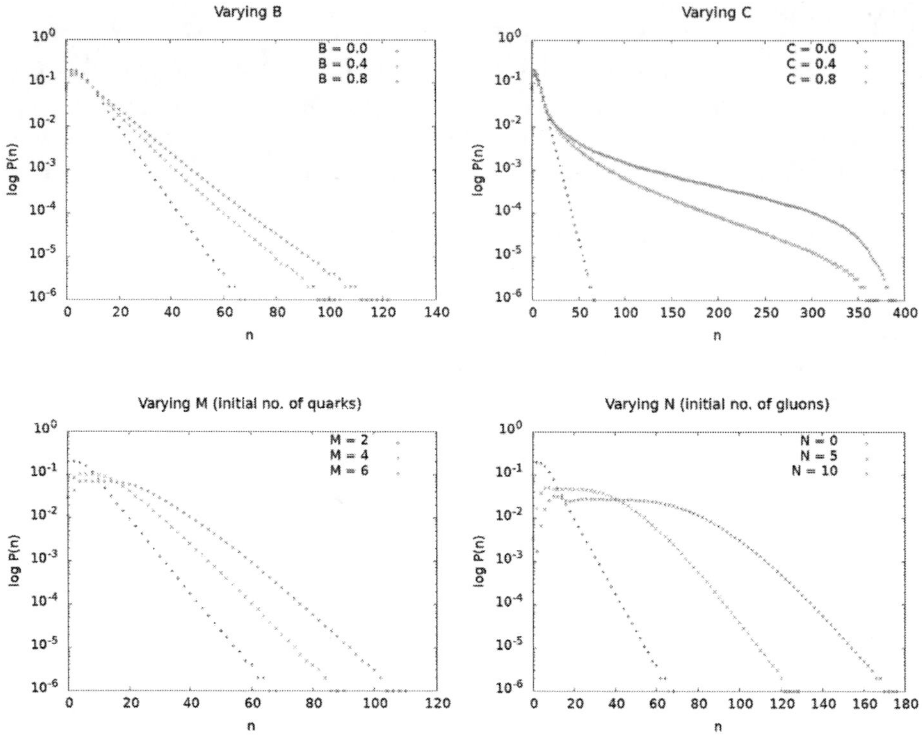

Summary of Observations

From the plots shown, we note the resulting effects of varying the six parameters:

1. Processes A ($g \to gg$) and C ($g \to ggg$) create the shoulder-like structure very strongly (a sharp change in gradient at approximately $n = 25$). This suggests that at onset of KNO-scaling violation, gluonic activity increases.

2. Processes $\widetilde{A}$ ($q \to qg$) and B ($g \to \overline{q}q$) generally broaden the entire distribution.

3. Higher values of M and N have the effect of shifting peak multiplicity to the right, while maintaining an overall bell shape of multiplicity distribution.

Conclusions

The emergence of gluonic activity (processes A and C) at the onset of KNO-scaling violation agrees with earlier predictions by Chan & Chew [2]. They observed that as $\sqrt{s}$ increased towards the TeV range, fits of collider data to the GMD revealed decreasing k and increasing k' (i.e. as gluon activity increases, distribution becomes more FYD).

In addition, we now have a method to observe theoretical distributions incorporating all four QCD processes, which also has the ability to reproduce the shoulder-like structure in KNO-scaling violation. This presents an alternate approach to the multicomponent-NBD/FYD/GMD models used to describe the shoulder-like structure.

Future Work

This preliminary work has hinted at a mechanism that possibly explains the onset of KNO-scaling violation, which warrants further investigation. We outline two areas of further development.

Fits to collider data: we can use the method outlined to perform best-fits for A, $\widetilde{A}$, B, C, M and N against multiplicity plots from collider data. This potentially reveals whether gluonic activity indeed increases with collision energy. We can also do the same for different types of collisions (e.g. $p\overline{p}$, $e^{+}e^{-}$, pA, AA).

Reduced phase space: So far, our model produces even-multiplicity plots, which correspond to full phase-space multiplicities. The model can be modified to produce multiplicity plots for all-n, which is more common today as colliders report charged particle multiplitICIES from reduced η-space.

Acknowledgements

The authors would like to thank the National University of Singapore, and the support and helpful discussions with colleagues. This work is supported by the NUS Research Scholarship.

References

[1] A. Giovannini, Nucl. Physics, Sect. B **161**, 429 (1979)
[2] A.H. Chan, C.K. Chew, Phys. Rev. D (1990)
[3] A. Dewanto, A.H. Chan, C.H. Oh, R. Chen, K. Sitaram, Eur. Phys. J. C **57**, 515 (2008)
[4] UA5 collaboration, Phys. Lett. B (1986)
[5] S. Sakai, Phys. Rev. D (1989)

Nonlinear Dirac Neutrino Oscillations

Zhi Hao Quek[1,*], *Wei Khim* Ng[1], *Aik Hui* Chan[1], and *Choo Hiap* Oh[1]

[1]National University of Singapore
Department of Physics
21 Lower Kent Ridge Rd, Singapore 119077

Abstract. Neutrino oscillations are a possible way to probe beyond Standard Model physics. The propagation of Dirac neutrinos in a massive medium is governed by the Dirac equation modified with an effective Hamiltonian that depends on the number density of surrounding matter fields. At the same time, quantum nonlinearities may contribute to neutrino oscillations by further modifying the Dirac equation. A possible nonlinearity is computationally studied using Mathematica at low energies. We find that the presence of a uniform, static background matter distribution may significantly alter the oscillation amplitude and wavelength; the considered nonlinearity may further reduce both oscillation amplitude and wavelength. In addition, the presence of matter allows the effects of the nonlinearity to be more readily observed for the chosen background densities and neutrino energy.

1 Introduction

Neutrino oscillations are largely regarded as quantum phenomena arising from the mixing between mass and flavour eigenstates. In the presence of surrounding matter, neutrino oscillations would be affected by charged and neutral current interactions.

At the same time, neutrino phenomena present interesting avenues for probing non-standard physics, such as quantum nonlinearities. These nonlinearities may be crafted via information-theoretic considerations, and function as effective theories; such nonlinearities may manifest as modifications to the effective equations of motion; alternatively, they may occur at the level of the Lagrangian.

In this paper, we consider both matter effects and quantum nonlinearities using a computational model.

2 Neutrino Oscillations Theory

Consider the following Lagrangian:

$$\mathscr{L} = \mathscr{L}_0 + \mathscr{L}_{\text{Matter}} + \mathscr{L}_{\text{NL}} \tag{1}$$

where $\mathscr{L}_0 = \bar{\nu}(i\gamma^\mu\partial_\mu - m)\nu$ is the usual free Dirac Lagrangian for neutral, spin-$\frac{1}{2}$ particles. $\mathscr{L}_{\text{Matter}}$ and $\mathscr{L}_{\text{NL}}$ represent the matter interaction and nonlinearity terms respectively.

*e-mail: zhihao.quek16@sps.nus.edu.sg

2.1 Matter Interaction

In standard matter, neutrinos can interact with background matter via charged and neutral current interactions:

$$\mathscr{L}_{\text{Matter}} = -\sqrt{2}G_F\bar{\nu}\gamma^\mu[J_\mu^{CC}N^{CC} + J_\mu^{NC}N^{NC}]P_L\nu \tag{2}$$

In the flavour basis, and for standard matter (whereby electron number density dominates over muon and tau number densities), $N^{CC} = \text{diag}(1,0,0)$ while $N^{NC} = I_{3\times3}$. Assuming that the background matter is static and unpolarized, that is $J_\mu \propto \rho(x)\,\delta_{\mu0}$, we obtain the following equations of motion [1]:

$$(i\gamma^\mu\partial_\mu - V_{\text{int}}\gamma^0)\nu_L - m\nu_R = 0 \tag{3a}$$

$$(i\gamma^\mu\partial_\mu)\nu_R - m\nu_L = 0 \tag{3b}$$

Here, $V_{\text{int}} = -2\,G_F\sum_f(T_{3_f} - 2Q_f\sin^2\theta_W)\,\rho_f\,N^{NC} + 2\,\rho_e\,N^{CC}$, where f represents different fermionic matter fields, T and Q are their isospin, charge and θ_W is the Weinberg angle, and ρ_f are the number densities of the various fermionic fields.

We further adopt the chiral representation for the γ matrices: $\gamma^0 = \begin{pmatrix} 0 & I \\ I & 0 \end{pmatrix}$ and $\gamma^i = \begin{pmatrix} 0 & \sigma^i \\ -\sigma^i & 0 \end{pmatrix}$.

The right-handed field in Equation 3 may be eliminated, as per references [2] and [3], to obtain a 2nd order equation of motion, which, for energy eigenstates, has the following conserved quantity:

$$N \equiv \nu_L^\dagger(-i\overleftrightarrow{\nabla} - \vec{\sigma}A)\nu_L \tag{4}$$

Subsequently, we use the above conserved quantity to normalize our probability distribution; this gives us the following expression for (electron) neutrino survival probability:

$$P_{e\to e} = \frac{1}{N}\nu_{eL}^\dagger(-i\overleftrightarrow{\nabla} - \vec{\sigma}A)\nu_{eL} \tag{5}$$

2.2 Quantum Nonlinearities

As per reference [4] we now proceed to introduce nonlinearities at the level of the Lagrangian. Here, we consider the following nonlinearity:

$$\mathcal{L}_{NL} = \frac{(i\bar{\nu}\gamma^\kappa\partial_\kappa\nu - i(\partial_\kappa\bar{\nu})\gamma^\kappa\nu)^2}{4A_\mu\bar{\nu}\gamma^\mu\nu} \tag{6}$$

Here, $A_\mu = (\frac{1}{A},0,0,0)$ is a constant timelike vector that results in Lorentz violation and A controls the strength of the nonlinearity. The four-dimensional integral of such a nonlinearity is posited to be an information measure; concomitantly, this nonlinearity satisfies the following information-theoretic axioms:

Firstly, the nonlinearity is homogeneous under re-scaling of ν, that is $\mathcal{L}_{NL}(\lambda\bar{\nu},\lambda\nu) = \lambda^2\mathcal{L}_{NL}$.

Secondly, the nonlinearity (and consequently the associated information measure) decrease in value as $\nu(t,\vec{x})$ tends to a constant value. This is guaranteed by the derivative term in the numerator of the nonlinearity, and corresponds to the intuition that spatial information of a system is lost when the state vector of the system tends to a constant (in the spatial representation).

Thirdly, the theory is local and thus only depends on $\nu, \bar{\nu}$ and their derivatives at a point.

Fourthly, the information measure must be positive for arbitrary ν.

Finally, the information measure is minimal when the nonlinearity is extremized to obtain the equations of motion.

From the above nonlinearity, we obtain the following modification to the equations of motion:

$$\begin{aligned}\partial_\mu \frac{\partial \mathcal{L}_{NL}}{\partial(\partial\bar{\nu})} - \frac{\partial \mathcal{L}_{NL}}{\partial \bar{\nu}} &= i\gamma^\mu \nu\, \partial_\mu \left[\frac{(i\bar{\nu}\gamma^\kappa \partial_\kappa \nu - i(\partial_\kappa \bar{\nu})\gamma^\kappa \nu)}{2A_\sigma \bar{\nu}\gamma^\sigma \nu} \right] - A_\mu \gamma^\mu \nu \left[\frac{(\bar{\nu}\gamma^\kappa \partial_\kappa \nu - (\partial_\kappa \bar{\nu})\gamma^\kappa \nu)}{2A_\sigma \bar{\nu}\gamma^\sigma \nu} \right]^2 \\ &\approx i\gamma^\mu \nu\, \partial_\mu \left[\frac{m\bar{\nu}\nu}{A_\sigma \bar{\nu}\gamma^\sigma \nu} \right] + A_\mu \gamma^\mu \nu \left[\frac{m\bar{\nu}\nu}{A_\sigma \bar{\nu}\gamma^\sigma \nu} \right]^2 \\ &\approx A_\mu \gamma^\mu \nu \left[\frac{m\bar{\nu}\nu}{A_\sigma \bar{\nu}\gamma^\sigma \nu} \right]^2 \end{aligned} \tag{7}$$

The first approximation arises from a perturbative calculation using the 0th order Dirac equation. This allows us to sidestep additional boundary conditions and by preventing the equation of motion from becoming a 2nd order PDE. The second approximation uses the fact that the derivatives of the plane-wave solution to the 0th order Dirac equation are null.

3 Methodology

Equations 3 and 7 combined give the effective propagation equation for Dirac neutrinos in matter. This EOM is studied numerically in Mathematica via a 4th order Runge Kutta method. The neutrinos are assumed to propagate along 1 dimension only, and are further assumed to be in a negative spin eigenstate, that is, $\sigma^x \nu = -\nu$. Furthermore, the neutrinos are assumed to be in an energy eigenstate, such that $\nu(t, \vec{x}) \propto e^{-iEt}$. Finally, the neutrinos are chosen to be initially in the electron flavor eigenstate.

4 Results

For ease of discussion, the results are presented in the following manner: matter interaction only, nonlinearity only, and finally both matter and nonlinearity.

4.1 Matter Interaction Only

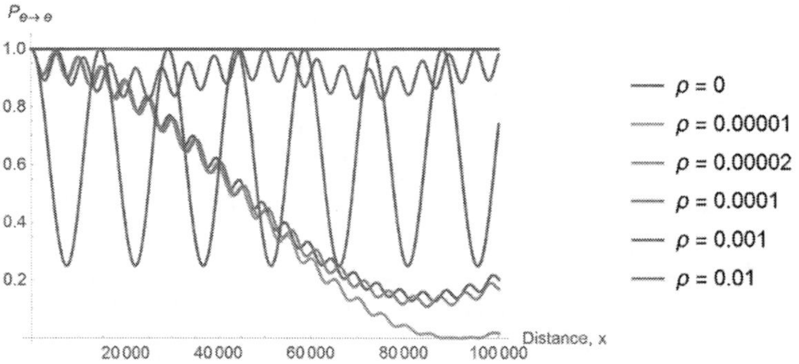

Figure 1. Plot of electron neutrino survival probability, $P_{e\to e}$, as a function of distance, x, for different background densities.

Here, standard matter is assumed (comprised by protons, neutrons and electrons with equal number densities). In general, the electron neutrino survival probability depends strongly on the background number densities, with high number densities resulting an almost complete suppression of the oscillation phenomena.

4.2 Nonlinearity Only

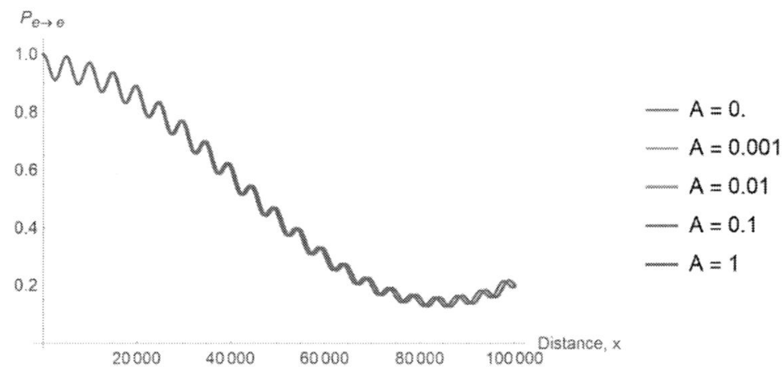

Figure 2. Plot of electron neutrino survival probability, $P_{e\to e}$, as a function of distance, x, for various nonlinearity strengths (different values of A). A larger A value indicates a stronger nonlinearity.

For the chosen range of nonlinearity strengths, we observe that the nonlinearity term does not significantly affect the oscillation amplitude; even with the largest nonlinearity of $A = 1$, we see only a slight shift in the oscillation wavelength with no significant change in amplitude.

4.3 Both Matter Interaction and Nonlinearity

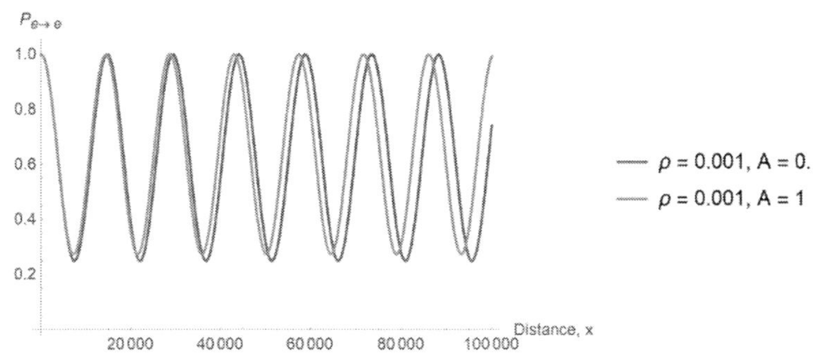

Figure 3. Plot of electron neutrino survival probability, $P_{e\to e}$, as a function of distance, x, for various nonlinearity strengths (different values of A), for nonzero background matter number density. A larger A value indicates a stronger nonlinearity.

With a nonzero background matter density, we see a much stronger effect with the introduction of the nonlinearity. In particular, we note a noticeable reduction in both wavelength and amplitude. This is in contrast with the much smaller change in oscillation amplitude shown in nonlinearity only case, i.e. Figure 2.

https://doi.org/10.1051/epjconf/202024007010

5 Conclusion and Future Works

We implement a modified equation of motion for Dirac particles via quantum nonlinearity to the case of neutrino oscillations in matter. Through our computation results, we see that it is in principle possible to use the phenomena of neutrino oscillations to probe quantum nonlinearities, wherein a sufficiently strong nonlinearity might manifest in a modifications to the survival probability (as a function of distance).

Further studies may explore how the oscillation probability varies as a function of energy. This may prove insightful in determining both the optimal energy for experimental probes of quantum nonlinearities and for placing upper-bounds on nonlinearity values by experiments with fixed detector distance.

6 Acknowledgements

The authors would like to thank organizing committee of the Southeast Asian Astronomy Network (SEAAN) Meeting 2019 for a wonderful conference and members of the NUS High Energy Physics group for meaningful discussions.

References

[1] T. K. Kuo and J. Pantaleone, Rev. Mod. Phys. **61:4**, (1989)
[2] Z. Y. Law, A. H. Chan, and C. H. Oh, Int. J. Mod. Phys. A **24**, 3483–3492 (2009)
[3] H. T. He, Z. Y. Law, A. H. Chan, and C. H. Oh, in Proc. Conf. Honour Murray Gell-Mann's 80th Birthd. Quantum Mech. Elem. Part. Quantum Cosmol. Complex, (2011)
[4] W. K. Ng and R. R. Parwani, Mod. Phys. Lett. A **26:9**, 681-689 (2011)

Closing up onto our sun: solar imaging

Kushagra Shrivastava[1*], *Keith Wen Kai* Chia [2], *Kang Jun* Wong[1], *Alfred Yong Liang Tan*[1] and *Hwee Tiang* Ning[1]

[1]National Junior College, 37 Hillcrest Road, Singapore
[2]School of Science and Technology, 1 Technology Drive, Singapore

Abstract. Solar activity research provides insight into the Sun's past, future (Science Daily, 2018). The solar activity includes observations of large numbers of intense sunspots, flares, and other phenomena; and demands a wide range of techniques and measurements on the observations. This research needs long term data collection before critical analyses can occur, to generate meaningful learning and knowledge. In this project, we will use solar imaging to make observations of solar activity, and take our baby steps to make contributions in citizen science. Observations will be made in 3 wavelengths to gain a more thorough analysis by looking at different perspectives of the Sun, namely H-Alpha, Calcium-K, and white light.

1. Introduction

In today's world, we rely heavily on satellites, radio communication and other such technology. They have made our life so much easier and convenient. Can you even imagine your life without YouTube, Instagram or Netflix? However, our closest star, the Sun, is a threat to this technology. Even though the Sun is the source of thermal energy on earth which makes it habitable for us, it is a double-edged sword. Solar activities like solar flares, coronal mass ejections and coronal holes bombard our satellites with charged particles that disrupt their normal functioning [1]. On March 13 1989, Quebec suffered a blackout as a result of a solar flare [2]. Due to such incidents, it is necessary for us to study about the Sun in order to predict when a solar activity might occur. Unfortunately, even after decades, we are not completely able to predict these events. Thus, everyone's observation is required – both amateur and professional solar astronomers – in order for us to predict when such events might take place.

2. Aim

The aim of our project is to document the activities on the Sun via optical means and analyze them in order to obtain more knowledge about the Sun and its activities.

3. Methodology

3.1 Equipment used

3.1.1 Hardware

- 2 Go2Nova telescope mount controllers (See Figure. 1)

*Email: kushagrashrivastava22@gmail.com

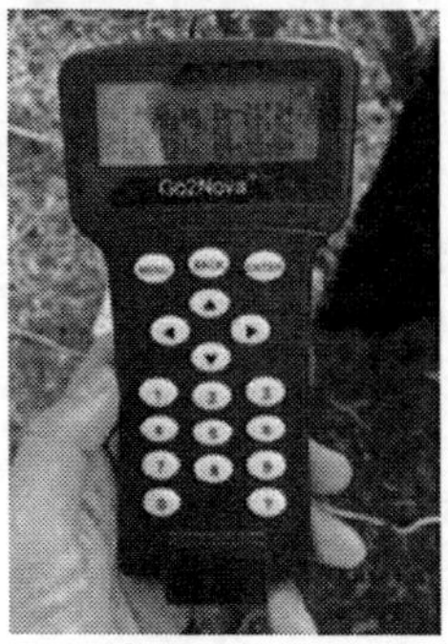

Figure 1. Go2Pro Controller

- CEM25P and CEM60 mount (See Figure 2)

Figure 2. Borg 76 mm refractor mounted on CEM25P mount (left) and CEM60 mount (right)

- Borg 76 mm achromat refractor (See Figure 2 left)
- Megrez-90 refractor (See Figure 3)

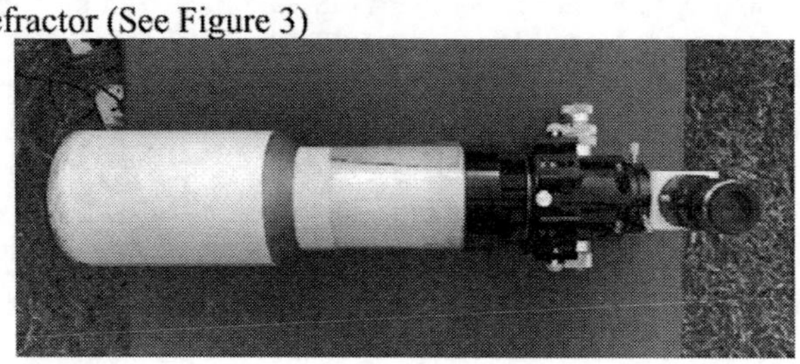

Figure 3. Megrez-90 Refractor

- White Light filter (See Figure 4)

Figure 4. White Light Filter

- Calcium-K filter (See Figure 5)

Figure 5. Calcium-K Filter

- Double stacked H-alpha Etalon (See Figure 6)

Figure 6. H-Alpha etalon

- ASI174MM Camera (See Figure 7)

Figure 7. ASI174MM

3.1.2 Software

- FireCapture
- AutoStakkert3!
- ImPPG
- Adobe Photoshop

3.2 Setting up mounts

The mounts were setup using the following procedure:

i. Place the mount on a flat ground

ii. Attach counterweight
iii. Perform polar alignment to the geographical North Pole
iv. Place telescope on the mount rails
v. Plug the Go2Nova controller onto the mount

3.3 Data capturing

For data capturing, the following procedure was used:

i. Using the Go2Nova controllers connected to the CEM25P and CEM60 separately, the telescope was aimed at the Sun.
ii. The camera, ASI174MM, was attached to a laptop with FireCapture.
iii. The mount controller was then used for fine adjustments and to bring the Sun in the center of the frame.
iv. Exposure was increased to see the three mirrored images of the Sun (See Figure 8), followed by decreasing it so that only the brightest one is visible.

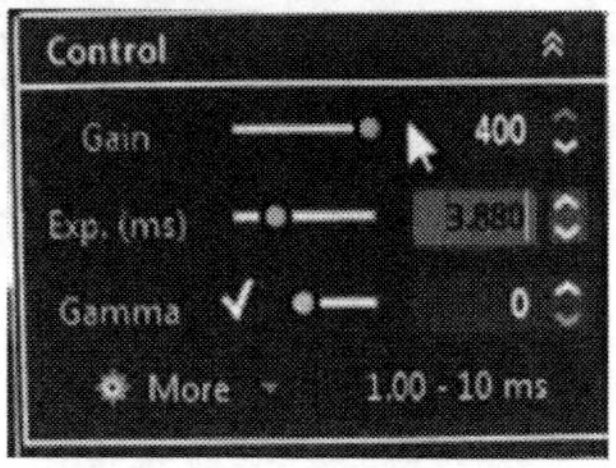

Figure 8. Increasing Exposure on FireCapture

v. The gain was increased or decreased as required.
vi. The video was captured with a time limit of 15 seconds.

3.4 Data processing

The problem with data processing is that there is no fixed set of parameters that can be used to process all images. It largely depends on the weather conditions and the quality of equipment used. In order to determine the best set of values for the various variables, we tried many and varied permutations of values, over many hours of processing. The values mentioned in the procedure below are the most used ones. However, it should be noted that every image requires slightly different parameters for fine-tuning. Thus, it is possible that the parameters mentioned afterward will not yield a good image for other users.

3.4.1 AutoStakkert3!

The recorded video was stacked using AutoStakkert3! First, the frames of the video were analyzed to obtain a frame quality curve of the video at surface image stabilization. Laplace transformation was on with noise-robust of 4 units. Then the frame quality curve was analyzed to see the percentage of frames with good quality followed by setting of alignment points at 16-unit size. 40-50% of the total frames were chosen for stacking. The sharpening box was unchecked and the video was stacked in TIF format. (See Figure 9)

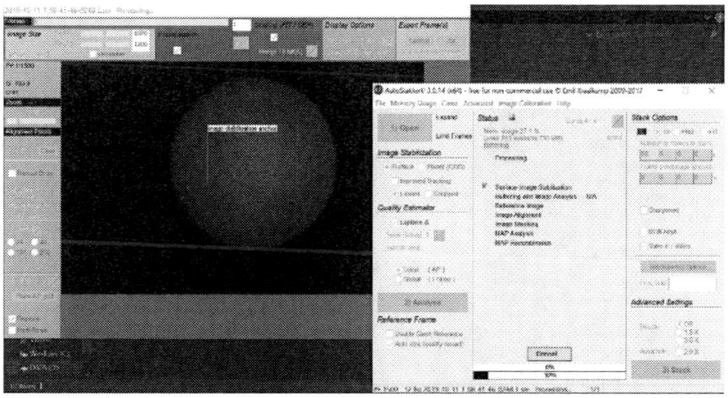

Figure 9. AutoStakkert3! analyzing a video

3.4.2 ImPPG

The stacked image was sharpened in ImPPG. The tone curve of the image is stretched to fit the histogram in order to get a higher contrast (See Figure 10). Then, the Lucy-Richardson deconvolution method [6] was used to sharpen the image. The values of sigma, amount and unsharp-masking were changed as per the image quality. Usually, the value of sigma and unsharp-masking was about the same. The final picture was then saved as a BMP file.

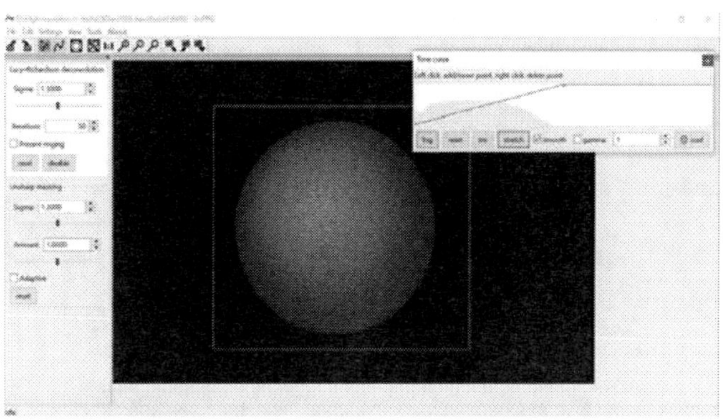

Figure 10. Stretching a tone curve in ImPPG

3.4.3 Adobe Photoshop

The sharpened image was then colored using the RGB tone curves in Adobe Photoshop. The color of the image was dependent on the layer of the Sun. Red-orangish colour was used for the upper chromosphere viewed using H-Alpha. Purplish-blue colour was used for lower chromosphere viewed using Calcium-K. Dark green colour was used for the photosphere viewed using white light. Then, an additional orientation step was used on some images to rotate the image so that the x-axis of the image corresponds to the north and the y-axis corresponds to the east. The orientation was done with respect to images from other solar observatories [2,3]. This was done in order to track solar features.

Figure 11. Colouring a H-Alpha image in Adobe Photoshop

4. Results and discussion

4.1 Hydrogen-Alpha Images

The images obtained with the Hydrogen-Alpha filter were monochrome, as a black and white camera was used. A reddish-orange pseudo-colour was added to the image (See Figure 12) as this colour will be produced by the filter if viewed through the eyepiece because we are looking at the upper chromosphere which emits a wavelength of 656.3 nm [7], which corresponds to this colour on the visible spectrum. Upon careful observation, an active region can be seen towards the south of the solar disk, which might develop into a sunspot. It has a streak like appearance and is brighter than the solar disk. However, this active region did not develop into a sunspot. Figure 13, shows a close-up view of the active region.

It should be noted that Figure 13 is not a high-resolution image and just a zoomed-in image. High-resolution imaging was not carried out in this research.

Figure 12. Upper Chromosphere seen via Hydrogen Alpha

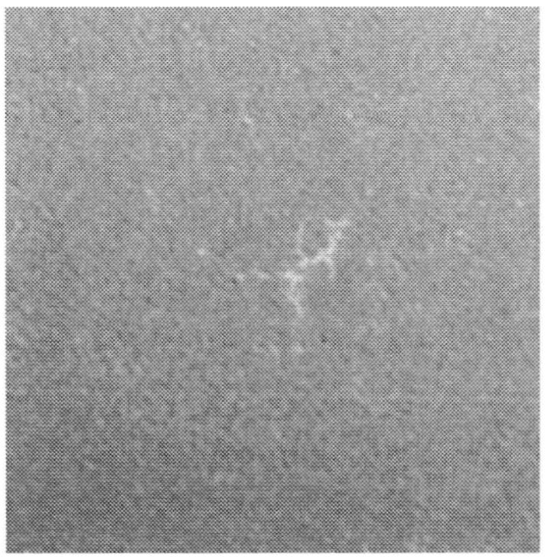

Figure 13. Active Region seen in H-Alpha

4.2 Calcium-K Images

Similar to H-Alpha, the Calcium-K images were monochrome. A purplish-blue pseudo-colour was added (See Figure 14) as colour close to this will be produced if seen through the eyepiece because it is the lower chromosphere and it emits wavelength of 393.4 nm [8] that corresponds to this colour on the visible spectrum. Figure 15 shows a close-up view of the solar limb. Again, it should be noted that it is not a high-resolution image. Unfortunately, no prominent solar activity is visible on the date of capture.

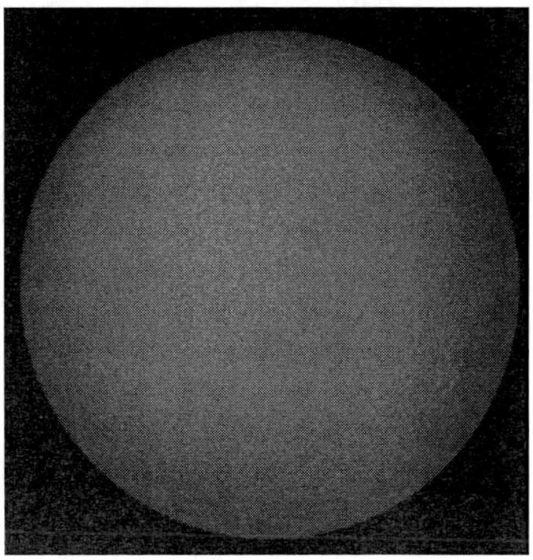

Figure 14. Lower Chromosphere seen via Calcium-K

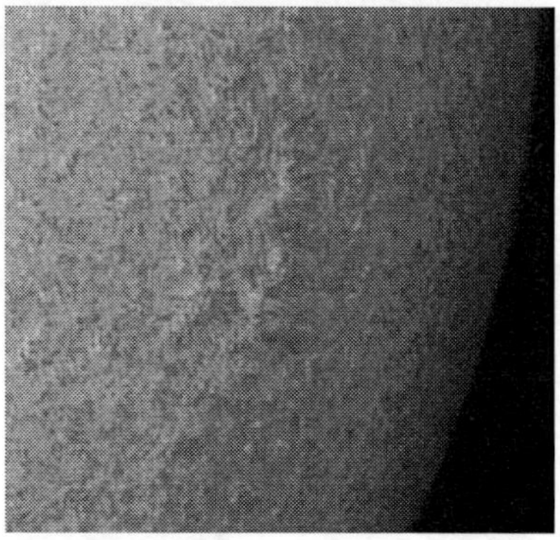

Figure 15. Close-up of Solar Limb in Calcium-K

4.3 White Light Images

Like the other two filters, the White Light images were monochrome as well. A dark green pseudo-colour was added to the image as colour close to this will be produced if seen through the eyepiece. This is the surface of the Sun, the photosphere, and we cannot see beyond this layer [9]. It can be seen that in Figure 16, no prominent solar activity of any kind is observed and the Sun appears inactive.

Figure 16. Photosphere seen through White Light

5. Conclusion

Not much solar activity can be seen on the Sun. The obtained images are very "plain" and "inactive". This is due to the fact that the Sun follows a 11-year activity cycle [5]. During the period of data collection, the Sun was at its solar minima, i.e., the period of minimum activity.

6. Future Works

In the future, we are starting on high-resolution imaging to focus on certain parts of the Sun rather than observing the solar disk as a whole. We have already obtained some preliminary results for high-resolution imaging. Figure 17 shows a high-resolution image of a sunspot. The difference in contrast around the sunspot is clearly visible. Also, patterns formed by the magnetic field lines can be seen as well.

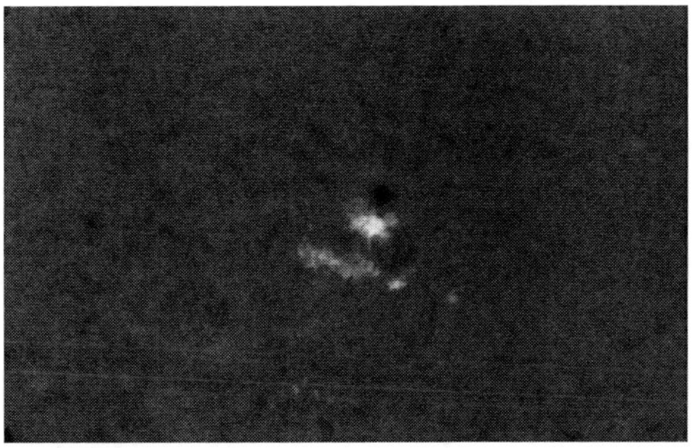

Figure 17. High-Resolution Image of a sunspot

Apart from high-resolution imaging, we might consider using the G-band filter as well to observe the Sun in a different wavelength.

7. Safety Concerns

- Handling of telescopes should be done with care as small shocks can cause the lenses to disorientate or damage the etalon.
- During data collection, it should be ensured that the filters are fitted on the telescope before any direct observation is made to ensure no harm is done to the eye.
- Proper sun protection gear – sunscreen, sunglasses, hats, et cetera – is used to prevent sunburns.

8. Acknowledgements

We would like to express our utmost gratitude to our research mentors, Ms. Ning Hwee Tiang and Mr. Alfred Tan, of National Junior College, for their continuous support and guidance throughout this research project. We would like to give special thanks to Mr. Alfred Tan, for letting us use his equipment and resources. We would also like to thank our schools, National Junior College and School of Science and Technology, for giving us the opportunity of this research project.

8. References

1. Sunspots and Solar Flares | NASA Space Place – NASA Science for Kids. (2019). Spaceplace.nasa.gov. Retrieved 25 December 2019, from https://spaceplace.nasa.govsolar-activity/en/

2. Addison, K. (2019). SDO | Solar Dynamics Observatory. Sdo.gsfc.nasa.gov. Retrieved 25 December 2019, from https://sdo.gsfc.nasa.gov/data/

3. Sun today — latest images of the Sun. (2019). Tesis.lebedev.ru. Retrieved 25 December 2019, from https://tesis.lebedev.ru/en/Sun_pictures.html

4. D. Hathway, *The Solar Cycle. Living Reviews In Solar Physics*, **12**, 1 (2015)

5. Solar experts predict the Sun's activity in Solar Cycle 25 to be below average, similar to Solar Cycle 24. (2019). Weather.gov. Retrieved 25 December 2019, from https://www.weather.gov/news/190504-Sun-activity-in-solar-cycle

6. Szczerek, F. (2020). ImPPG Processing Tutorial. Retrieved from https://greatattractor.github.io/ImPPG/tutorial/tutorial_en.html

7. Hydrogen Alpha Explained. (2020). Retrieved from http://www.astronomyknowhow.com/hydrogen-alpha.htm

8. 7 - Observing the Sun in Ca-K, Ca-H and Other Narrow Bandwidths | AstronomyConnect. (2020). Retrieved from https://astronomyconnect.com/forums/articles/7-observing-the-Sun-in-ca-k-ca-h-and-other-narrow-bandwidths.38/

9. The Photosphere - the "Surface" of the Sun | UCAR Center for Science Education. (2020). Retrieved from https://scied.ucar.edu/Sun-photosphere

EPJ Web of Conferences **240**, 07012 (2020) https://doi.org/10.1051/epjconf/202024007012
SEAAN Meeting 2019

Search for Stellar Streams in The Galactic Halo From Gaia DR2, GALAH DR2, RAVE DR5 and LAMOST DR4 Data

Ni Made Kartika Wijayanti[1,*], *Mochamad* Ikbal Arifyanto[2,**], and *Nur* Annisa[1]

[1]Astronomy Study Program, Faculty of Mathematics and Natural Sciences, Institut Teknologi Bandung, Indonesia

[2]Astronomy Research Division, Faculty of Mathematics and Natural Sciences, Institut Teknologi Bandung, Indonesia

Abstract. Stellar streams are stars which are trapped in the same potensial caused by dynamical resonance or tidal force. We aim to analyze kinematic substructures (streams) in the Galactic halo by V vs $\sqrt{U^2 + 2V^2}$ planes of Arifyanto & Fuchs. We crossmatched data from Gaia DR2, GALAH DR2, RAVE DR5 and LAMOST DR4 based on positions. We have 3D kinematics and metallicity data of halo stars selected from kinematics criteria from ratio of probability of thick disk (TD) over halo (H) less than 0.01. Substructures are detected by using wavelet transformation and corrected using 15 Monte Carlo simulations. We obtained four kinematic structures on V vs $\sqrt{U^2 + 2V^2}$ plane which two of them are associated to BB17-1 and BB17-2 streams. All the streams had a high probability from the extragalactic origin.

1 Introduction

There are two scenarios of galactic formation. First is monolithic collapse by Eggen, Lynden-Bell, and Sandage [1]. In monolithic collapse, a galaxy is formed from the same protogalaxy cloud which collapsed and every part of galaxy, such as bulge, disk, and halo, is formed at the same time. The second one is hierarchical scenario [2]. In this scenario, each part of galaxy is formed from different clouds then merge into one and forming a galaxy. This scenario can explain that stars in bulge, disk, and halo have different typical age. Hierarchical scenario is supported by a proof that Sagittarius dwarf spheroidal (Sgr dSph) deformed because of tidal force from Galactic center [3]. Some globular clusters such as M54, Arp2, Terzan7, and Terzan8 are suggested that came from Sgr dSph. Helmi and White [4] also showed that local stellar halo is builded by disrupted satellites.

Eggen introduced "moving group" as stars from an open cluster or from the same protostar cloud. The stars then escaped from the system so they are detected as a group of stars with the same velocity [5]. In 1978, Eggen introduced "retrograde group," stars that expected came from ω Centauri cluster [6]. Metalicity and age of stars in retrograde group are broad in range so this suggests that ω Centauri was a galaxy and stripped into Milky Way [7].

There are two types of streams: dynamical stream and tidal steam [8]. Stars that trapped into the same dynamical resonance, such as bar resonance and spiral arm resonance, is called

*e-mail: kartika.wijayanti@students.itb.ac.id
**e-mail: ikbal@as.itb.ac.id

dynamical stream. Stars that came from the same bounded object, such as satellite galaxy, can escaped from the system and enter the Milky Way forming tidal stream. Search for stellar streams are using star grouping in velocities space or integral of motion space. Dehnen [9], Skuljan [10], and Kushniruk et al. [11] use U vs V space for streams in Galactic disk. Several well-known streams are Hyades, Pleiades, and Sirius. These streams were thought to be originated from disrupted open cluster but their isochrone showed possibility of dynamical resonant origin [12]. Arifyanto & Fuchs [13] and Klement [8] use V vs $\sqrt{U^2+2V^2}$ space to identify streams in disk. Bajkova and Bobylev [14] use V vs $\sqrt{U^2+2V^2}$ space for searching streams in Galactic halo. Another integral of motion spaces for searching streams are V_{az} vs $V_{\Delta E}$ ([15], [16]), L_z vs $L_\perp$ [8], and L_z vs E ([17], [18]).

In this work, stellar streams are searched using V vs $\sqrt{U^2+2V^2}$ planes. U is velocity vector towards Galactic center, V is velocity vector in the direction of rotation of Galaxy and $\sqrt{U^2+2V^2}$ is related to eccentricity of star's explained in Section 3.

2 Data

To get velocities, we use parallax and proper motion data from Gaia DR2 [19] and radial velocity data from LAMOST DR4 [20], RAVE DR5 [21], and GALAH DR2 [22]. LAMOST DR4, RAVE DR5, and GALAH DR2 are crossmatched into Gaia DR2 data based on star's position with 0.8" radius. We select solar neighborhood stars with galactocentric radius within 7 to 10 kpc and distance from galactic plane $Z < 5$ kpc. We also eliminate stars with total velocity more than 580 kms^{-1} which is the Galactic escape velocity [23]. The selected stars are transformed to UVW velocities using matrix from [24] as shown in equation 1.

$$\begin{bmatrix} U \\ V \\ W \end{bmatrix} = \mathbf{T}.\begin{bmatrix} +\cos\alpha\cos\delta & -\sin\alpha & -\cos\alpha\sin\delta \\ +\sin\alpha\cos\delta & +\cos\alpha & -\sin\alpha\sin\delta \\ +\sin\delta & 0 & +\cos\delta \end{bmatrix}\begin{bmatrix} V_r \\ k\mu_\alpha/\varpi \\ k\mu_\delta/\varpi \end{bmatrix} + \begin{bmatrix} U_\odot \\ V_\odot \\ W_\odot \end{bmatrix}, \tag{1}$$

where $\mathbf{T}$ is transformation matrix from galactic coordinates to equatorial coordinates, α and δ are equatorial coordinates, V_r is radial velocity, k is 4,74057 to transform all the units into kms^{-1}, μ_α and μ_δ are proper motion in equatorial coordinates, and ϖ is parallax of the stars. We use Sun's velocity component is $(U, V, W)_\odot = (11.1, 12.24, 7.25)$ kms^{-1} [25].

Galactic halo stars are selected kinematically as performed in [26]. We use probability of thick disk over halo $TD/H = \frac{X_{TD}}{X_H}\frac{f_{TD}}{f_H} < 0.01$ to reduce contamination from thick disk. Each X_{TD} and X_H is fraction number of thick disk and halo respectively. Each f_{TD} and f_H is Gaussian distribution function for stars in thick disk and halo respectively. The function of Gaussian distribution of 3D stellar kinematics is

$$f = \frac{1}{(2\pi)^{3/2}\sigma_U\sigma_V\sigma_W}\exp - \frac{U^2}{2\sigma_U^2} - \frac{(V-V_{asym})^2}{2\sigma_V^2} - \frac{W^2}{2\sigma_W^2}, \tag{2}$$

where σ_U, σ_V, and σ_W is velocity dispersions of each stellar population and V_{asym} is asymmetric drift relative to LSR. Value of σ_U, σ_V, σ_W, and V_{asym} in equation 2 is shown in Table 1.

3 Search Method

To find streams, we identify groupings on V vs $\sqrt{U^2+2V^2}$ space. Value of $\sqrt{U^2+2V^2}$ represents eccentricity of the star. Using Taylor expansion around $1/R$ and $1/R_0$ until second

Table 1: Fraction number and characteristics of Gaussian distribution for stars in thick disk and halo.

	X	σ_U	σ_U	σ_U	V_{asym}
		kms^{-1}			
Thick Disk (TD)	0.09	67	38	35	-46
Halo (H)	0.0015	160	90	90	-220

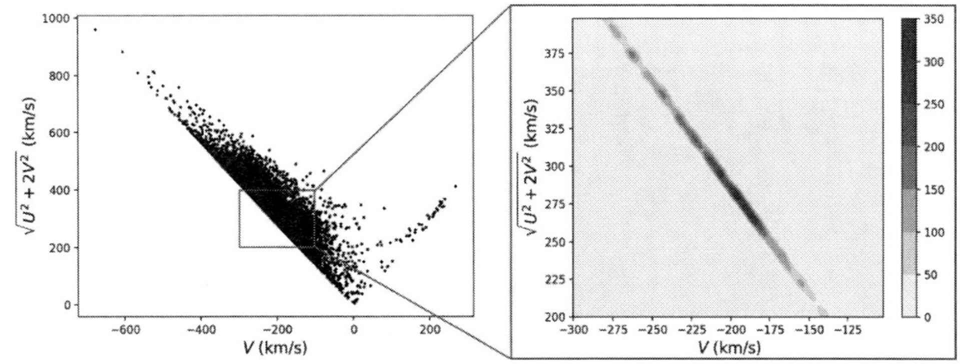

Fig. 1. Sample data distribution on V vs $\sqrt{U^2+2V^2}$ space (*left panel*). Wavelet transformation of the data in range of the red box is shown on the *right panel*.

order, Dekker's definition of angular velocity $\Omega(R) = \sqrt{\frac{1}{R}\frac{d\Phi}{dR}}$ [27], and energy of stars in the guiding center $E_0 = \Phi(R_0) + \frac{1}{2}R_0^2\Omega^2(R_0)$, Arifyanto & Fuchs [13] got eccentricity of stars is

$$e = \sqrt{\frac{2(E-E_0)}{R_0^2\kappa_0^2}} = \sqrt{\frac{U^2 + \frac{\kappa_0^2}{\Omega_0^2}V^2}{R_0^2\kappa_0^2}}. \tag{3}$$

For flat rotation curve, $\frac{\kappa_0^2}{\Omega_0^2} = 2$ and $R_0^2\kappa_0^2 = 2V^2$ so $e \propto \sqrt{U^2+2V^2}$.

Distribution of the data on V vs $\sqrt{U^2+2V^2}$ space is shown on left panel of Fig. 1. The groupings on this distribution are found using Mexican-hat wavelet transform from Skuljan [10]. Wavelet is a powerful way to extract signal from data distribution. We divide the data into several bins and calculate the wavelet coefficient of each bin using equation 4. For a two-dimensional distribution $f(x,y)$ at any point (ξ,η), the function of wavelet coefficient is

$$w(x,y) = \int_{-\infty}^{\infty}\int_{-\infty}^{\infty} f(x,y)\psi\Big(\frac{x-\xi}{a}, \frac{y-\eta}{a}\Big)dxdy, \tag{4}$$

where ψ is the analysing wavelet function. Here we use two-dimesional Mexican hat $\psi(d/a) = (2-(d/a)^2)\exp(-d^2/(2a^2))$, where $d = \sqrt{x^2+y^2}$ or the distance from bin to the data. The bin is square with 2 kms^{-1} width and scale parameter a is 4 kms^{-1}. The result of the transformation is shown in right panel of Fig. 1. We show significant groupings in range $V = [-300, -100]$ kms^{-1} and $\sqrt{U^2+2V^2} = [200, 400]$ kms^{-1} because otherwise is relatively smooth.

The detected groupings are corrected using Monte Carlo simulation to reduce the statistic fluctuation. We generate 15 simulations, each simulation has the same number of stars as

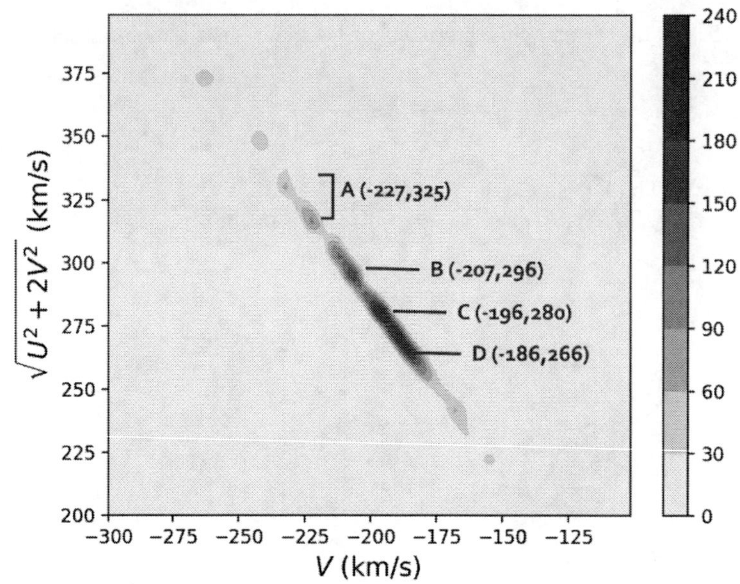

Fig. 2. Four groupings found after cleaning the data using Monte Carlo simulation.

our data. The synthetic data from simulation have the same Gaussian distribution as Galactic halo. Each simulation is treated the same as the data to get the wavelet transformation with the same size of bin. Each simulation is subtracted with transformed data. The mean of each substraction is shown in Fig. 2.

Table 2: Streams found and associated streams.

Streams found	Coordinates kms^{-1}	Associated streams	Coordinates of associated streams kms^{-1}
A	(-186,266)	BB17-1 [14]	(-170,250)
B	(-196,280)		
C	(-207,296)		
D	(-227,325)	BB17-2 [14]	(-220, 330)

4 Results

In this work, we found four streams or groupings as shown in Fig. 2 and Table 2. We labeled the groupings as A, B, C, and D. We suggest that two of them, A and D, are associated to BB17-1 and BB17-2, high velocity streams in Galactic halo by Bajkova & Bobylev [14]. Bajkova & Bobylev found three streams named BB17-1, BB17-2, and BB17-3. BB17-1 and BB17-2 are correlated to VelHel-6 and VelHel-7 streams by Helmi et al. [17]. Several streams found by Helmi et al., named VelHel-1, 2, 3, 4, 5, 8, and 9, have positions close to ω Centauri globular cluster. Properties of this cluster support the idea of ω Centauri probably was a galaxy and stripped into Milky Way. ω Centauri have a high mass ($\sim 7 \times 10^6$ $M_\odot$), flattened shape or tidal features, wide range of abundance and age, and retrograde orbit [7]. Other streams in Galactic halo may be the result of dwarf satellite galaxy. On the next work, we will analyze more about origin of these streams.

Acknowledgement

The authors present results from the European Space Agency (ESA) space mission Gaia (https://www.cosmos.esa.int/gaia), processed by the Gaia Data Processing and Analysis Consortium (DPAC). We also used data from GALAH (https://galah-survey.org/), RAVE (https://www.rave-survey.org), and Guoshoujing Telescope the Large Sky Area Multi-Object Fiber Spectroscopic Telescope LAMOST (http://dr4.lamost.org/).

References

[1] O.J. Eggen, D. Lynden-Bell, A.R. Sandage, ApJ **136**, 748 (1962).
[2] L. Searle, R. Zinn, ApJ **225**, 357 (1978).
[3] R.A. Ibata, G. Gilmore, M.J. Irwin, Nature **370**, 194-196 (1994).
[4] A. Helmi and S.D.M. White, MNRAS **307**, 495-517 (1999).
[5] O.J. Eggen, AJ **112**, 1595 (1996).
[6] O.J. Eggen, ApJ **221**, 881-892 (1978).
[7] S.R. Majewski, R.J. Patterson, D.I. Dinescu, W.Y. Johnson, J.C. Ostheimer, W.E. Kunkel, C. Palma, arXiv 9910278 (1999).
[8] R. Klement, B. Fuchs, H.W. Rix, ApJ **685**, 261-271 (2008).
[9] W. Dehnen, AAS **115**, 2384 (1998).
[10] J. Skuljan, J.B. Hearnshaw, P.L. Cottrell, MNRAS **308**, 731-740 (1999).
[11] I. Kushniruk, T. Schirmer, T. Bensby, A&A **608**, A73 (2017).
[12] B. Famaey, A. Siebert, A. Jorissen, A&A **483**, 453-459 (2008).
[13] M.I. Arifyanto & B. Fuchs, A&A **449**, 533-538 (2006).
[14] A.T. Bajkova & V.V. Bobylev, Astron. Lett. **44**, 193-201 (2018).
[15] C. Dettbarn, B. Fuchs, C. Flynn, M. Williams, A&A **474**, 857-861 (2007).
[16] R.J.Klement, Astron Astrophys Rev **18**, 567-594 (2010).
[17] A. Helmi, et al. 2017. A&A, 598, A58
[18] H. Li, C. Du, S. Liu, T. Donlon, H.J. Newberg, ApJ **874**, 1 (2019).
[19] Gaia Collaboration, A&A **616**, A1 (2018a).
[20] A.L. Luo, Y.H. Zhao, G. Zhao, VizieR Online Data Catalog, V-153 (2018).
[21] RAVE Collaboration. AJ **153**, 75 (2017).
[22] GALAH Collaboration, MNRAS **478**, 4513-4552 (2018).
[23] Monari, G. et al . 2018. A&A 616, L9.
[24] D.R. Johnson & D.R. Soderblom, AJ **93**, 864-867 (1986).
[25] R. Schönrich, J. Binney, and Walter Dehnen
[26] T. Bensby, S. Feltzing, M.S. Oey, A&A **562**, A71 (2014).
[27] E. Dekker, Phys. Rep. **24**, 5 (1976).

AUTHOR INDEX

EDP Sciences – Web of Conferences
17, Avenue du Hoggar
Parc d'Activité de Courtabouf
BP 112
F-91944 Les Ulis Cedex A
France

ISBN 978-1-7138-1796-3

CURRAN ASSOCIATES INC.
proceedings.com

3rd International Conference on Chemical and Food Engineering 2016

MATEC Web of Conferences
Volume 62 (2016)

Tokyo, Japan
8-9 April 2016

edp sciences